The History of Environmental Degradation in Mar Menor

This book offers a multidisciplinary analysis of the degradation process of an ecosystem, drawing upon the Mar Menor as a case study to highlight the damage human pressure causes to the environment.

All ecosystems change over time, although in some cases, this variation is more dynamic and evident. The Mar Menor is a clear example of this "ecological transition", as it is the largest coastal lagoon in the western Mediterranean and the first ecosystem in Europe to be granted legal personhood rights. This book provides an extensive overview of the history of its environmental degradation over the past 100 years, highlighting the subsequent succession of environmental crises including phytoplankton explosions, the disappearance of large areas of submerged meadows due to eutrophication, and episodes of mass mortality of aquatic fauna. Split into three sections to reflect thematic blocks, the book begins with a comprehensive description of the Mar Menor and its marine ecosystems, emphasizing its ecological value and unique space in Spain and Europe. It discusses intensive and globalized agriculture, surrounding agro-export, and the laws that legislate it. In the second part, the book draws on a series of cultural concepts, theoretical frameworks, and participatory arts-based research to enrich our understanding of the environment from multiple perspectives. Finally, in the third part, the book uses analysis gathered from the Mar Menor case study to discuss wider conclusions about the ways in which we can begin to undo our damage to the environment and restore ecosystems.

This book will be useful for students, academics, and researchers interested in environmental justice, environmental history and anthropology, sustainable development, and environmental studies more broadly.

Juan Manuel Zaragoza is Postdoctoral Researcher at the University of Murcia (Spain). From 2013 to 2015, he was a Marie Curie Research Fellow at the Centre for the History of the Emotions at Queen Mary University of London, and from 2015 to 2016 a BBVA Foundation Leonardo Fellow. His research has focused on the history of experience and emotions, specifically around the experiences of discomfort and well-being. He is PI of the project *Climate crisis, mental health and well-being in the Anthropocene* and Founding Director of the research collective ehCOLAB (https://www.um.es/ehcolab/) interested in the development of the blue humanities.

David Soto Carrasco is Full Professor of Moral Philosophy at the University of Murcia and Head of the Department of Philosophy at the same university. He has also been Full Professor at the Universidad del Pacífico in Ecuador. His work focuses on the study of the philosophical and political dimensions of the crisis of the 1930s and 1970s, especially in Spain and Latin America. He has also worked on environmental ethics and environmental public policies in the Southeast Pacific, editing five collective books with the Universidad del Pacífico, the last one entitled *La Declaración de Santiago de 1952: una alianza del Pacífico Sudeste sobre políticas marítimas y ambientales*. Some of his most recent publications are *Filosofía política y ética: claves conceptuales para comprender el presente*; "Políticas del terror: subjetividad neoliberal y populismo autoritario", en: Turpín Saorín, J. (ed.). *Antropología en devenir político* (2023).

Malena Canteros graduated in Philosophy from the University of Murcia (Spain). She has completed a Master's degree in Philosophy Research with a specialization in Contemporary Aesthetics, as well as another Master's degree in Teacher Training, both at the University of Murcia. Her current area of interest is the relationship between climate change and mental health through a gender perspective.

Routledge Environmental History

The new *Routledge Environmental History* series presents current interdisciplinary research investigating our changing ecological entanglements over time. It draws on diverse time periods, geographical regions, methods and ideas from subject areas such as history, geography, archaeology, and the natural sciences. It taps into how topical debates such as the roles of non-humans and more-than-humans, social and environmental justice, climate change, and antiracism are shaping environmental histories and shifting approaches to engaging with the natural world.

Comprising edited collections, co-authored volumes and single author monographs, this collection provides an invaluable resource for advanced undergraduate and postgraduate students and scholars with an interest in the wide-ranging field of environmental history.

For more information, or to submit a proposal, please contact the Commissioning Editor, Grace Harrison (grace.harrison@tandf.co.uk).

A Global Environmental History of Coastal Dunes
Joana Gaspar de Freitas

How the California Electricity Crisis Generated a Green Wave
An Insider's Account
Kurt Schuparra

A Cultural History of Waste Disposal
Environmental Policy and Park Redevelopments
Benjamin A. Lawson

The History of Environmental Degradation in Mar Menor
A Case Study
Edited by Juan Manuel Zaragoza, David Soto Carrasco and Malena Canteros

For more information about this series, please visit: www.routledge.com/Routledge-Environmental-History/book-series/ESREH

The History of Environmental Degradation in Mar Menor

A Case Study

Edited by
Juan Manuel Zaragoza, David Soto Carrasco and Malena Canteros

First published 2025
by Routledge
4 Park Square, Milton Park, Abingdon, Oxon OX14 4RN

and by Routledge
605 Third Avenue, New York, NY 10158

Routledge is an imprint of the Taylor & Francis Group, an informa business

British Library Cataloguing-in-Publication Data
A catalogue record for this book is available from the British Library

ISBN: 978-1-032-78696-4 (hbk)
ISBN: 978-1-032-78697-1 (pbk)
ISBN: 978-1-003-48907-8 (ebk)

DOI: 10.4324/9781003489078

Typeset in Times New Roman
by codeMantra

Contents

Contributors

María José Alcaraz León is Associate Professor of Aesthetics at the University of Murcia where she teaches Contemporary Aesthetics, Philosophy of Literature, and Cultural Landscapes. Her main research interests are aesthetic normativity, art theory, fiction and cognitive value, art and morality, and environmental aesthetics. She has published in the *Journal of Aesthetics and Art Criticism*, *Estetika*: *The European Journal of Aesthetics*, *Teorema*, *Enrahonar*, and *Thémata*. In 2008, she was awarded the John Fisher Memorial Prize for her essay on the rational justification of aesthetic judgments and in 2013 she was awarded the Young Researches Prize by the Región de Murcia. During the course 2023–2024, she will be visiting research in Uppsala.

Alba Ballester is a facilitator, consultant, trainer, and researcher in water governance, public participation, and socio-institutional capacity building in natural hazards. Her work focuses on conflict transformation in hydrological risks. Her research interests are water governance, risk governance, and water conflict management. She has been teaching Social Anthropology at the University of Zaragoza, Spain. Her research has focused on the analysis of public participation on water planning, social capacity building in relation to floods management, the design of natural hazards prevention public policies, and the analysis of new forms of collective action.

Gustavo A. Ballesteros Pelegrín is Professor of Geography at the Autonomous University of Madrid. His scientific and professional activities have focused on the monitoring, planning, management, and conservation of wetlands and waterbirds in the Region of Murcia. He has led the process of inventorying, delineating, and processing three wetlands for designation as *Birds Special Protection Areas* (SPAs), two of which were also included as Ramsar sites. He has also coordinated the process for the inclusion of 53 wetlands in the Region of Murcia in the National Inventory of Wetlands and has managed two European Union LIFE projects: LIFE-MALVASIA and LIFE-SALINAS.

Isabel Banos-González is Spanish researcher in the field of ecology and sustainability. She holds a PhD in Ecology and Hydrology from the University of Murcia and has published numerous articles in scientific journals on topics such as sustainable water use, modeling socio-ecological systems, and environmental education. She is also co-author of several books on these topics. Banos-González has worked on various research projects related to sustainability in islands and Mediterranean areas. She is a member of several research groups and has participated in international conferences and congresses. Currently, Banos-González is Associate Professor at the University of Murcia. Her research areas focus on the analysis of sustainability in socio-ecological systems, the use of dynamic models for the evaluation of environmental policies, and education for sustainability.

Pedro Baños Páez obtained PhD in Sociology at the University of Murcia; degree in (i) Political Science and Sociology at the National University of Distance Education, (ii) Industrial Technical Engineer at the Polytechnic University of Cartagena; Diploma in Environmental Engineering in the School of Industrial Organization, Ministry of Industry and Energy; is Associate Professor in the Department of Sociology, University of Murcia. He is author/co-author of books and book chapters and articles published in national and international journals on the social and environmental problems of the Sierra Minera de La Unión-Cartagena and their impact on Portmán Bay and the Mar Menor. He has received several awards and public recognition from neighborhood and nature conservation associations for his research on the social and environmental problems of the Sierra Minera de La Unión-Cartagena and Portman Bay.

Mijo Miquel Bartual holds a degree in Philology and Fine Arts, PhD in Public Art (2013) and is a research member of the History and Philosophy of Experience group of the Center for Human and Social Sciences of the CSIC (2014–2023). She has been working in the field of languages (teaching, translation, and interpretation) from 1993 to the present. Since 2003, she is Professor of Sculpture at the Faculty of Fine Arts of San Carlos as well as an independent cultural manager. As a researcher, it is worth mentioning her constant involvement in seminars and various meetings as both organizer and speaker, as well as being the author of numerous scientific publications. She collaborates with different recognized university masters (Ecology, Urban Regeneration, or Art Therapy) and in research and innovation projects at European level.

Violeta Cabello is Research Fellow at the Basque Centre for Climate Change. She is a multidisciplinary environmental scientist with a special focus on knowledge co-production for complex social-ecological problems. She also works as facilitator of collective and participatory processes. She works within both academic and civil society organizations bridging different types of knowledge, connecting theory with practice and creating transdisciplinary networks across scales. Her research focuses on knowledge co-production for sustainability transformations. She has developed a two-year participatory research process in the Mar Menor context (2021–2022).

M. Francisca Carreño Fructuoso is Researcher at the Geological-Mining Technology Unit of the Technological Centre for Mineral Resources and Materials, Cehegín, Murcia, where she has been working for the last three years on research into multi- and hyperspectral remote sensing applied to mining exploration and rehabilitation. Her doctoral thesis focused on land use changes in the Mar Menor basin and their impact on communities and habitats, including coastal wetlands and their associated avifauna. He has been a member of the Mediterranean Ecosystems Research Group (University of Murcia) for more than 20 years and has been part of the research teams of the Sustainability Observatory of the Region of Murcia (2009–2012) and the Marine Sciences and Applied Biology Group of the University of Alicante (2018–2019).

Carlos de Castro Pericacho is Associate Professor in the Department of Sociology at the Universidad Autónoma of Madrid. He has been a Fulbright Visiting Professor at the University of California, San Diego. His work focuses on the study of the social, political, and cultural dimensions of the processes of insertion of productive territories in global production chains. Some of his most recent publications are co-editor and author of *La producción de la calidad en el sector agroalimentario: un análisis sociológico* (Tirant, 2022); "The Nature of Standards: How Standards Shape the Value of Nature". *International Sociology*, 37(6) (with Miguel Ángel Sánchez and Andrés Pedreño Cánovas); co-editor and co-author of *Migration and Agriculture: Mobility and Change in the Mediterranean Area* (Routledge, 2017); "Designer Grapes: the Socio-Technical Construction of the Seedless Table Grapes. A Case Study of Quality Control". *Sociologia Ruralis*, 58(2) 2018 (w/Cristóbal Torres) (c.decastro@uam.es).

María Giménez Casalduero is Lawyer and Associate Lecturer in the Department of Administrative Law at the University of Murcia and in the Department of State Legal Studies at the University of Alicante and holds a PhD in Law in 2024. As an environmental lawyer, she has worked for the International Institute of Law and Environment (IIDMA) based in Madrid and the Interamerican Association of Environmental Law (AIDA-Americas) based in La Paz (Baja California Sur-Mexico) and currently she maintains an active collaboration with the international environmental justice organization ClientEarth (based in London) advising as a lawyer on marine and fisheries environmental issues.

Encarna Guillén obtained his degree in Environmental Sciences with a thesis on the environmental indicator value of steppe birds in areas surrounding the Mar Menor lagoon.

Nicolás Gutiérrez is Research Assistant in research line 4 – Adaptation Line – at BC3 Basque Centre for Climate Change, as well as a PhD student in models and research areas in social sciences at the University of the Basque Country. His work focuses on the study of environmental conflicts, specifically on the production of identities and ecological embeddedness of the actors involved, supporting the promotion of participatory public policies on environmental issues. He has been a teacher of basic and secondary education and a researcher in development cooperation organizations. His favorite place to co-produce knowledge is natural spaces, currently the mountains of Leioa in the Basque Country.

Miguel Ángel Martínez works as a researcher, teacher, and curator. He did a Master's in Performing Arts (2009) and holds a PhD in Literary Studies from the Universitat de València (2016). He also did the Artists Program at the Universidad Torcuato Di Tella (Argentina, 2018). Between 2017 and 2019, he developed postdoctoral research attached to CONICET and UNTREF (Buenos Aires, Argentina), and between 2019 and 2021 one linked to the Universitat de València and the Universitat Autònoma de Barcelona. He is now Professor at the Universitat de València and in the "Articulacions" Program, the artistic studies program of IVAM (Valencian Institute of Modern Art), the Universitat de València, and the Universitat Politècnica de València. He has recently published the books *La otra fiesta* (IVAM, 2020) and *Bios: Literatura, enfermedad, formas de vida* (Tirant Lo Blanch, 2021).

Julio Mas Hernández (Cartagena, 1953) is Doctor in Biology from the Autonomous University of Madrid. The title of the doctoral thesis was "El Mar Menor. Relations, Differences and Affinities between the Coastal Lagoon and the adjacent Mediterranean Sea" (1994). Principal Investigator of the Spanish Institute of Oceanography (IEO), he has been Director of the Oceanographic Center of Murcia for more than a decade. He also developed his professional work in the Fisheries and Aquaculture Service of the Autonomous Community of the Region of Murcia. He has been Vice President of Greenpeace Spain, for more than a decade. He has collaborated since its inception with ANSE (Association of Naturalists of the Southeast) and with other entities and conservation NGOs, OCEANA, and among others. It is currently part of the Pact for the Mar Menor and collaborates with ADELA, PROCABO, and COLUMBARES. He has been awarded different prizes, among them the "Luis Ramírez Díaz", by the Official College of Biologists of the Region of Murcia.

Miguel Mesa del Castillo Clavel is Architect at the Escuela Técnica Superior de Arquitectura de Madrid and holds a PhD from the University of Alicante, where he is Full-Time Lecturer in the Architectural Projects Area since 2007. From 1996 to 2005, he lived in Rome (Italy) where he was research assistant for a year at the Università degli Studi "La Sapienza" in Rome. In 1998, after traveling to Tel-Aviv and Tulkarem (Palestine), he began to work on the political implications of architectural practices in the Palestinian-Israeli territorial conflict, which occupies part of his academic dedication, both in teaching and research. Since 2010, he has been conducting research on the relevance of the ordinary in architecture from a socio-technical perspective and its ecological and environmental implications. He has been Guest Professor at the Institut D'Arquitectura Avançada de Catalunya, at the European University of Valencia, at the Istituto Nazionale di Architettura (InArch, Rome), at LaBoral in Gijón, at Tabakalera, Donostia, and at the Official Master in Models and Areas of Research in Social Sciences of the University of the Basque Country. Over the last six years, he has carried out a series of teaching programs and research work at the University of Alicante that transfer research interests to the pedagogy of the project in some specific topics: a review of the theoretical tools with which the practice of design is faced (STS, Political Philosophy, Political Ecology, Literary Creation, research, and art criticism); openness to the use of new tools in the design process; attention to the multiple scales in which architectural realizations are developed.

Andrés Pedreño Cánovas is Reader in the Department of Sociology, University of Murcia from 2001, Lecturer in the Department of Sociology, University of Murcia from 1993, holds PhD in Sociology from the University of Murcia in 1998, is member of Research Networks of Agrarian Labour, Inequalities and Rurality of CLACSO, Director of the Department of Sociology at the University of Murcia, Editor of *Revista de Sociología histórica* (*Review of Historical Sociology*) at the University of Murcia. Some of his most recent publications are editor and author of *De cadenas, migrantes y jornaleros: los territorios rurales en las cadenas globales agroalimentarias*, Talasa, Madrid; "New geographies in the global production of table grape: inequality and local diversity, Ager", *Revista de estudios sobre despoblación y desarrollo rural*, no. 24, pp. 35–62; "Sobre el 'espíritu' de la calidad y la nueva racionalización de la producción de frutas y uvas en la Región de Murcia", *RES. Revista Española de Sociología, vol. 30*, no. 1.

Maria Ptqk is a curator and cultural researcher. Born in Bilbao in 1976, she holds a PhD in Artistic Research from the University of the Basque Country, a degree in Law and Economics, a Master of Advanced Studies in International Public Law from Paris II-Sorbonne and in Cultural Law from the UNED-Carlos III in Madrid, and a Master in Cultural Management from the University of Barcelona. Her work is based on the intersections between art, technoscience, and feminisms and she is a member of the advisory group of the publishing house consonni. She has curated the exhibitions *Soft Power* (Amarika Proiektua Project, 2009), *A propósito del Chthuluceno y sus especies compañeras* (Espace virtuel du Jeu de Paume, Paris, 2017), *Reset Mar Menor. Laboratorio de imaginarios para un paisaje en crisis* (CCC Valencia, 2020), *Ciencia fricción. Vida entre especies compañeras* (CCCB Barcelona, 2021 – Finalist of the Asociaciò Catalana de Crítica d'Art Awards & Azkuna Zentroa, Bilbao, 2022), and *Extinción Remota Detectada* (LABoral, Gijon, 2022). She is the current curator of the International Image Festival Getxophoto.

Francisco Robledano holds a PhD in Biology and is Professor of Ecology at the University of Murcia. For four years, he was responsible in the Autonomous Community of the Region of Murcia for the management of protected natural areas of coastal and inland wetlands, before returning to the university to continue studying waterbirds as bioindicators of environmental change, especially in the Mar Menor area. He currently coordinates the Master's degree in Protected Areas, Natural Resources, and Biodiversity and is Vice Principal of the Mare Nostrum International Campus-Port of Cartagena Authority Interuniversity Chair for the Environment.

Pablo Rodríguez Ros (Cartagena, 1990) holds a PhD in Marine Sciences (UPC, 2020) and a degree in Environmental Sciences (UMU, 2013). He has worked as Researcher in Ocean Sciences for over ten years, conducting research stays and scientific expeditions on all continents and oceans. He has served as an advisor on marine and environmental affairs in the 3rd Vice Presidency and the Ministry for Ecological Transition and the Demographic Challenge, specifically regarding the Mar Menor crisis. Currently, he is the coordinator of Marine Protected Areas at the Marilles Foundation and pursues the objective of protecting 30% of the Mediterranean Sea by 2030. He is the author of the books *El mar que muere* (Balduque, 2023) and *Argonauta* (Raspabook, 2020).

Eduardo Salazar Ortuño holds a degree and law degree from the University of Murcia, specializing in Environmental Law Consultancy and Litigation, and is a university specialist in mediation for legal professionals – Magister Legum from the University of Dresden, Germany. He is Guest Lecturer on the Master's degree in Environmental Law and Sustainability at the University of Alicante and Associate Lecturer in Administrative Law at the Law faculty of the University of Murcia. He was awarded PRATS CANUT Prize for the best thesis published in Environmental Law in 2019, by the Rovira i Virgili University of Tarragona. His publications include the book *El acceso a la justicia ambiental a partir del Convenio de Aarhus. Justicia ambiental de la transición ecológica*, Thomson Reuters-Aranzadi, Cizur Menor, 2019.

Miguel Ángel Sánchez García holds a degree in Sociology from the University of Murcia and a PhD in Sociology from the Complutense University of Madrid. Since 2019, he has been working as a researcher in the project "Quality governance in global agri-food chains" (AgriQuality). Previously, he has been research assistant in the research project "Threat and diversity in urban context" at the University of Tübingen (Germany). Her main research interests are rural sociology, food consumption, and the literary field.

Teresa Vicente Giménez (Lorca, 1962) is Professor of Philosophy of Law and Director of the Chair of Human Rights and Rights of Nature at the University of Murcia. After practicing law between 1987 and 1994, her research and academic publications have focused on ecological justice, social rights, legal feminism, and children's rights. In 2019, together with a group of jurists, scientists, and activists, she promoted the Popular Legislative Initiative to recognize the legal personality of the Mar Menor in the Region of Murcia and thus gave the lagoon its own rights. In October 2021, this initiative reached the half a million signatures necessary for its bill to be debated in the Congress of Deputies.

Antonio Zamora López is Pre-doctoral Researcher in the Department of Zoology and Physical Anthropology, University of Murcia. Although he is currently working on his PhD on the fish communities of the Mar Menor, he focused his final year and Master's thesis on the population trends and habitat selection of waterbirds in the lagoon. His enthusiasm for this group of birds and his training as a bander for more than a decade have motivated him to participate in various research projects on seabirds and larolimic populations. As a result of the knowledge acquired and the information generated with many collaborators, the book *Atlas de las aves acuáticas del Mar Menor y humedales de su entorno* has been published, of which he is a co-author. His interests and lines of research extend to other groups of birds, such as passerines and nocturnal birds of prey, contributing to studies assessing their response to changes in land use and environmental pollutants.

Foreword

Pablo Rodríguez Ros

The Mar Menor Coastal Lagoon and Its Three Greens

And so the Mar Menor turned green. It was 2016, and I had been working outside the Region of Murcia for several years. A few years earlier, in 2008, I began my environmental science studies at the University of Murcia. Rare was the subject that did not discuss the Mar Menor – in Ecology, due to the diversity of its ecosystems, iconic life forms, or invasive species; in Land Management, due to the hypertrophy in land use (and therefore, abuse) and the overexploitation of natural resources; in Hidrology, due to nitrate pollution in groundwater and surface waters; in Environmental Sociology, we were told about the importance of the Mar Menor for the culture of the society that inhabits it; in Economics, the benefits that wetlands provide to society; in Soil Science, the limestone crust present in the soil of the Campo de Cartagena. And so we could continue listing the scientifically proven causes of the current state of this ecosystem. The Mar Menor encompasses everything because numerous, too many, elements related to global change occur there.

All those professors and lecturers from the academic world taught us, from their respective fields, about the situation of the Mar Menor. Or rather, they warned us. I remember reflecting, concerned, on the urgency to act. Something had to be done. The prevailing sentiment in society was that it was not necessary because, after all, "nothing" was happening. At the same time, public officials told us that what was happening in the Mar Menor was due to sunscreens, that a lettuce field was more effective in combating climate change than a forest ecosystem, or that the clarity of the waters of the Mar Menor reflected its good conservation status. And this third argument, despite being technically incorrect, was the one that caused the entire house of cards of inaction to collapse: the opaque green of the Mar Menor was, therefore, a negative sign. And part of society woke up. Thus, alongside the traditional environmentalist entities, new ones were formed, and in the following years, there was a successive atomization of small activist groups that has not yet ended.

This "negative" greenness occurred to a greater extent as a consequence of another greenness that, however, generates a positive sensation in many citizens. During the decades leading up to 2016, a new sea was born and grew, but this one was on land. And it was also green. The landscape, once brownish, began to turn intense green. And it was socially celebrated. Intensive monoculture agriculture expanded throughout the Campo de Cartagena, as if giant buckets of lettuce-green paint had been poured from the surrounding mountains, coloring the land all the way to the shores of the Mar Menor. Lettuces are currently cultivated so close to the Mar Menor that if a fish were bold enough, it could leap out of the water and feed on them. It would be amusing if we didn't have scientific evidence indicating that the pollution emanating from the Mar Menor, that shifting green, has also reached the Mediterranean and threatens marine reserves and *Posidonia oceanica* meadows.

The third green is yet to come, and to achieve it, this book is indispensable. I refer to the green plan that the Mar Menor needs to become the sea we need. This plan must arise from dialogue among administrations, governments, civil society, businesses, and other stakeholders. However, it requires information, both scientific and nonscientific, to help achieve its best design. Thus, while the Mar Menor struggled under the weight of environmental degradation, academics and activists embarked on a journey to document its decline and glimpse paths toward renewal. And this book is the result.

The book begins with the description of the problems facing the Mar Menor. Julio Mas narrates the evolution of the Mar Menor over time, while Francisco Robledano, Encarna Guillén, M. Francisca Carreño Fructuoso, Antonio Zamora López, and Gustavo A. Ballesteros Pelegrín showcase their results after investigating the biodiversity of its borders, using birds as indicators of environmental change. However, Andrés Pedreño Cánovas delves into intensive agriculture in the Campo de Cartagena, that "green blanket" we spoke of, unraveling the agricultural rationale and deep history that have shaped the region. Meanwhile, Carlos de Castro, Andrés Pedreño Cánovas, and Miguel Ángel Sánchez-Rodriguez examine the ecological degradation of the Mar Menor, highlighting the role of power, science, and deep histories in global agriculture. Finally, Pedro Baños and Isabel Banos-González explore the complexity of the Sierra Minera, revealing the challenges and opportunities facing this unique landscape in the fight for environmental conservation of the Mar Menor and its surroundings.

The second part of the book points to society and how it can and must be an unavoidable part of the solution. María José Alcaraz León delves into cultural landscapes, like those depicted by Carmen Conde, lamenting the loss of the aesthetic beauty that once defined the region. Meanwhile, Maria Ptqk showcases practices within the Chthulucene, offering a glimpse into innovative approaches to healing our wounded planet that H.P. Lovecraft would be proud of. Amidst the crisis, Violeta Cabello and Alba Ballester propose (and put to the test) a citizens' assembly as a beacon of hope for the eco-social transition of the Mar Menor and the Campo de Cartagena. This vision that the solution must come from society is shared by Teresa Vicente, who documents the legal recognition of the rights of nature of the Mar Menor, a crucial moment in the struggle for environmental justice. Amidst these narratives of struggle and resistance, María Giménez poses an inevitable question: is restoration possible after the collapse of the Mar Menor? It is a question that remains in the minds of all who witness the ecological tragedy unfolding in this ecosystem that once thrived.

This book ends with an intersection between culture and politics. Miguel Mesa del Castillo explores the intersection of reality and fiction in La Manga del Mar Menor, offering ordinary considerations for an exceptional environmental culture. Meanwhile, Mijo Miquel invites us to contemplate the assembly of the gaze, a powerful tool for understanding our relationship with the environment. Amidst the turmoil, Miguel Ángel Martínez shares stories of eco-social transition, highlighting moments of resistance and adaptation in the face of adversity. In turn, David Soto Carrasco's eco-socialist perspective on the crisis of the Mar Menor sheds light on the systemic injustices that perpetuate ecological degradation. And finally, Juan Manuel Zaragoza navigates the realm of minor emotions and offers a vision of the profound human experience of living in a damaged environment, a sentiment shared by all who call the Mar Menor home.

I deeply trust that this book will contribute to the Mar Menor and its people regaining the dignity and hope that have been taken from them. Only through joint and interdisciplinary work will society be provided with the necessary tools to achieve this. Likewise, I am convinced that this book will help us understand that without the environment, there is nothing: neither economy, nor culture, nor democracy – but also no well-being for present and future generations. Let us flee from weak consensus and work to make them strong, as strong as the commitment of the authors of this book. The third green is closer.

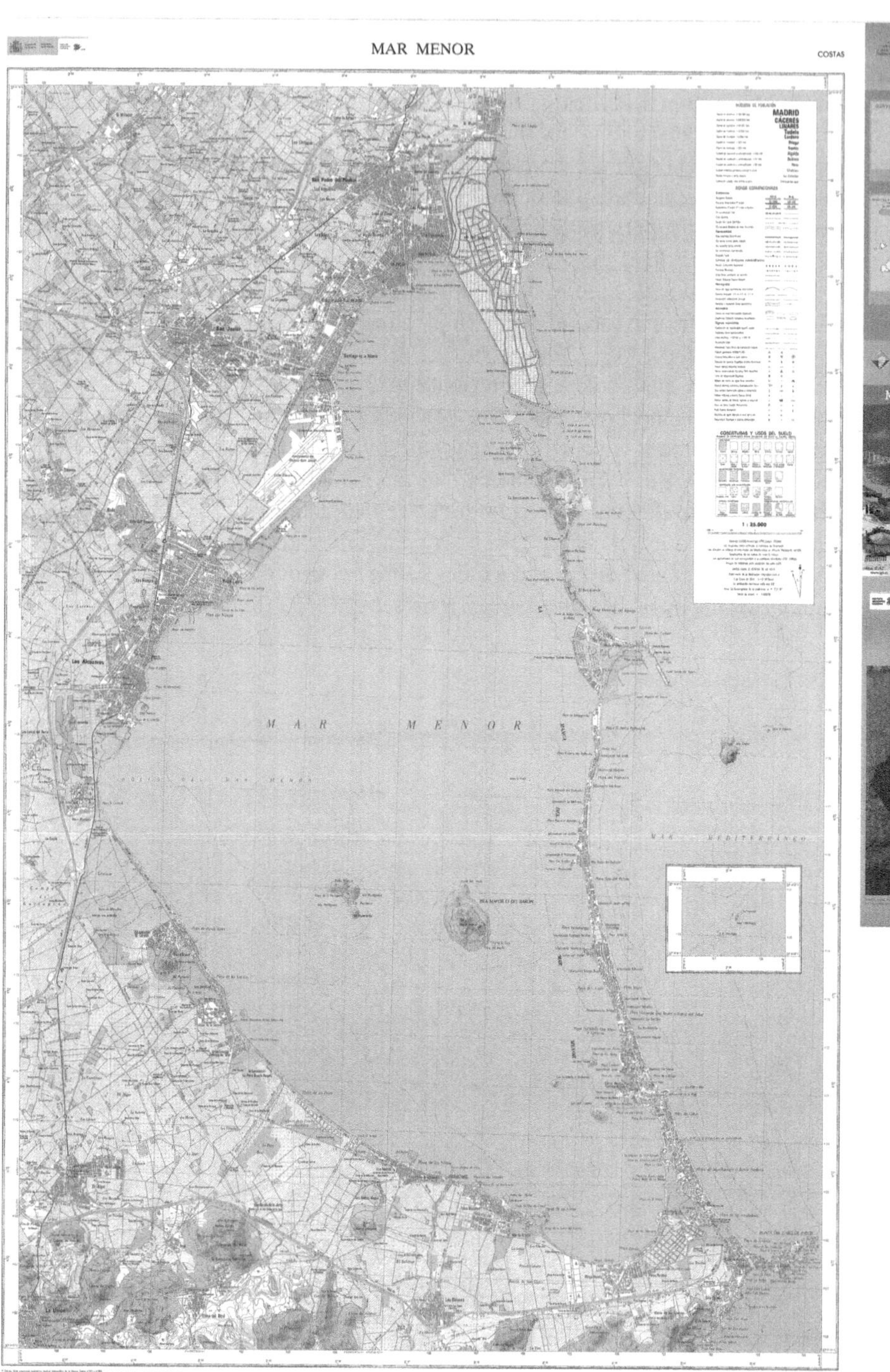
MAR MENOR
COSTAS
M A R   M E N O R
MAR MEDITERRANEO
1 : 25.000
Costas
Mar Menor

Introduction

Juan Manuel Zaragoza, David Soto Carrasco and Malena Canteros

"The crabs were trying to escape over the rocks, the shrimps were piling up, almost forming a paste, the eels were jumping, the water was boiling…" (Sánchez, 2019). This is how an eyewitness described what happened on the beaches of the Mar Menor on October 12, 2019 for an article published in El País – a sea that was "boiling", agitated by the efforts of hundreds and thousands of creatures to escape what had become a death trap. There are few things more terrifying than watching your home become uninhabitable – a place marked by death and destruction. The losses from this episode of anoxia[1] were immense, incalculable and were felt throughout the lagoon's surroundings: houses were devalued, fishing became unprofitable, tourists stopped coming. The economic impact was partially mitigated in the COVID crisis that would begin the following winter, but the consequences are long-lasting and many of them have not yet been overcome.

But this disaster of 2019 is not an event without history, on the contrary. It has a history and it is deeply intertwined with the human history of the configuration of a territory, the Campo de Cartagena (see Figure 0.1), with the emergence of a new political entity in post-Franco democratic Spain (the Autonomous Community of Murcia), and it is part of a long and profound process of modernisation in Spain, which began in the 19th century with the various "desamortizaciones",[2] was reinforced by the Regenerationist movement at the beginning of the 20th century,[3] and has continued throughout the century, regardless of the political regime, up to the present day. A fundamental part of this process of modernisation was to bring unproductive land into production, either by transforming it into irrigated land – for which an ambitious public works plan would be launched (Camprubi, 2017), of which the Tajo-Segura transfer would be one of the most outstanding (Morales Gil et al., 2005) – or those areas close to the sea, such as La Manga, which offered suitable conditions to be part of the incipient national tourist offer. This book you are holding in your hands is intended to give an account of this history of ecological degradation, which, as we shall see, has much to do with the (unequal, unjust, unbalanced) development of this modernisation project.

However, the analysis we propose here did not begin as a book, but as an exercise in activism, taking advantage of the – increasingly narrow – margins of manoeuvre that academia allows us. Ecological activism, of course, but also activism in defence of the public university, which we understood, could not remain on

DOI: 10.4324/9781003489078-1

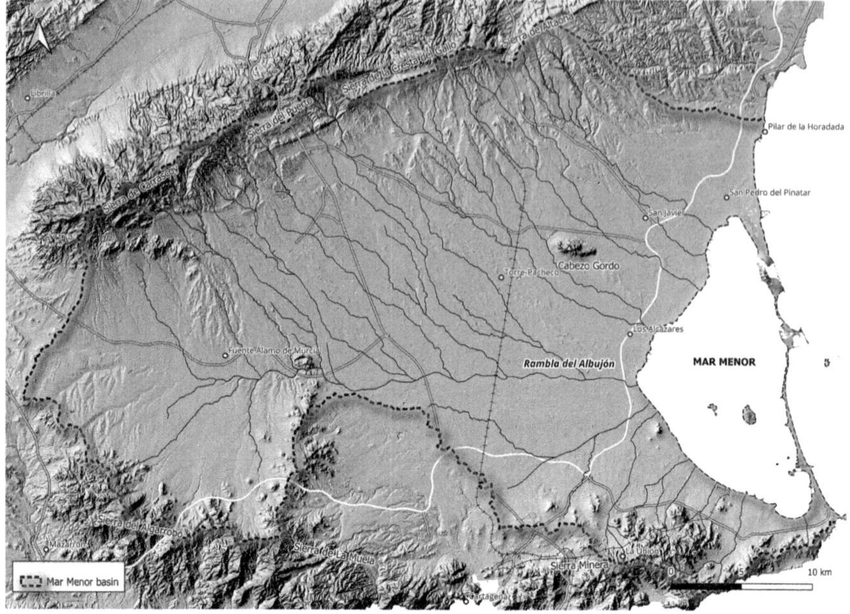

Figure 0.1 Map of the Mar Menor catchment area. 2024. Author: Alfredo Pérez Morales. University of Murcia.

the sidelines of the greatest eco-social crisis in our region. This is why a group of people linked in one way or another to the University of Murcia, through the collective Pensamiento al Margen (editors of the magazine of the same name), decided to organise a course to transfer the knowledge generated around the lagoon to Murcian society, in order to make sense of what was happening. The decisive factor in this process, which will mark the essence of the course and therefore of the book, is the composition of the Pensamiento al Margen collective, which is made up entirely of researchers from the humanities and social sciences. The challenge was therefore all the greater. It was a matter of making oneself heard where no one expected to hear one's voice. What could a group of philosophers, linguists and historians contribute to the debate on the state and recovery of the Mar Menor? What could we say?

The Mar Menor as a matter of concern

Not much, if we understand that the problem of the Mar Menor is exclusively related to the physico-chemical conditions of the lagoon – the presence of chemical products, the dynamics of the currents, the exchange of water with the Mediterranean, the strength of the food web and the balance of the ecosystem. But it is also true that nothing we said could ignore this material reality. The levels of oxygen, salt and phosphorus have been tirelessly measured and discussed ad nauseam. Also the dynamics of the currents or the resilience of the ecosystem. The limits of the discipline squeezed us like a second- or third-hand suit.

The Mar Menor destroyed everything that we considered a "fact" in the sense of disciplinary affiliation. Was it a "scientific fact"? Or should we think of it as a "social fact"? Or perhaps an "artistic representation"? A "political event" perhaps? The answer, at some point in our discussions, became clear. All of them are true: we were facing what Bruno Latour called "a matter of concern" (Latour, 2004b, 2008). What does this mean? In the first place, it means that we need to increase our zoom. To explain what is happening (to account for the totality of our experience, paraphrasing William James), we must take into account much more than what is at first sight. To use a theatrical metaphor, we need to move from observing only the scene to thinking about what happens backstage. That is why our observations on the sociology of agricultural work, on the deep history of the Region of Murcia, on the cultural and aesthetic values of the landscape, and on our emotional ties with the lagoon suddenly became relevant to understanding what had happened. Not in opposition to what the scientists were telling us, not by illuminating "other aspects of the problem", but by helping us to compose it, to try to describe it in its entirety. In this way, we constituted a new ecology of practices, a cosmopolitan space (Stengers, 2010) in which, on the basis of always local and precarious agreements and commitments, we felt able to make our Mar Menor emerge.

It was not simply a "change of focus", nor was it simply a matter of expanding the number of disciplines and fields of knowledge we needed to convene in order to obtain a more accurate description of what had happened. Moving from the matter of fact to the matter of concern, as Latour suggests, also implied the possibility of fostering open dialogue and collaboration between the various groups involved. That is, it led us to the task of expanding the political community far beyond what our modern standard allows, creating new assemblies, new parliaments (Latour, 2018). This conviction led us to include in our course aspects related to political participation and representation, to the point that our main objective was to offer the civil community tools to facilitate their participation in / organisation of decision-making processes in contexts of eco-social crisis.

This book shares both its orientation and its structure, opting for a multidisciplinary vision of the problem that should ultimately serve as part of citizen participation processes that allow complex decisions to be made from informed positions, deepening an idea of democracy as an experiment that we find both in the classical pragmatists – John Dewey, in this case (Honneth, 1998) – and in the aforementioned Bruno Latour (2004a). It was precisely this emphasis on citizen participation that attracted the attention of the European Climate Foundation, which supported us, first through Rosa Martínez Rodríguez – current Secretary of State for Social Rights of the Government of Spain – and later through Rosa Llobregat, in the successive implementation of two other projects that deepened the part focused on citizen participation. We would like to thank them for this support, which continues to this day.

The book is divided into three parts. The first, entitled "Describe", contains the contributions that help us to focus on what has happened (and in some cases is still happening) in the Mar Menor. To this end, as we have already indicated, we have tried to include all the voices that had something to contribute, regardless of their disciplinary origin. Thus, the first chapter, signed by the oceanographer Julio Mas, offers us a great overview of the history of the Mar Menor as an ecosystem, with special attention to the last 100 years, during which the lagoon has undergone its most profound changes, with significant and lasting ecosystem changes that have irreversibly altered the living conditions of the lagoon. The history of the Mar Menor offered by Professor Mas begins in the Pleistocene epoch, 18,000 years ago, with the formation of the sandbar that isolated the waters of the Mar Menor from those of the Mediterranean, and ends in the present day with the mention of the great white spot that stretches from the mouth of the Rambla del Albujón to the islands of Perdiguera and Barón, and which is currently the subject of the most intense scientific debate.

In the second chapter, the team led by Francisco Robledano takes us to the transitional areas between the Mar Menor and the surrounding area. These wetlands, which used to be common on the inner shores of the lagoon, used to mitigate the effects of agriculture on the Mar Menor, acting as natural filters and providing a habitat for a diverse fauna, especially waterfowl, many of which are in danger of extinction. However, the pressures of agriculture and urban development have led to a deterioration in the conditions of these areas, to the point, as the chapter argues, that the evolution of these ecosystems and the wildlife populations associated with them can become an excellent indicator for assessing the evolution of the Mar Menor as a whole. The authors therefore insist that the Mar Menor is not only the space occupied by the sheet of water, but that it cannot be separated from the environment in which it is located, which is why taking care of the Mar Menor ecosystem means intervening in the entire basin and not just in its immediate surroundings.

The third chapter takes us into the deep history of Campo de Cartagena in an attempt to understand the roots of the environmental destruction of the territory, which Pedreño, Castro and Sánchez identify with what they call the "agrarian reason". Inspired by Gramsci's work *The Southern Question* (Gramsci, 2005), they also seek to understand the current situation of the Mar Menor by looking at the historical roots of the agro-capitalist development of the southern peninsula and, more specifically, the Campo de Cartagena area. A development that was supported by a narrative called for a hydraulic revolution to transform the desert into a garden, which, in the hands of Francoist propaganda, soon became a patriotic and moral task. If it was necessary to "liberate" the nation from the drought that threatened it, the peasant's duty was to contribute to its eradication.

The fourth chapter, signed by the same authors, takes us back to the agricultural world, but this time from a different perspective, looking at how the controversy over the causes of the Green Soup events of 2016 and the massive fish kills of 2019 and 2021 has been produced, in what is identified as an attempt to focus the debate on these extremes, avoiding discussion of the measures that should be implemented. The authors point out that we are facing a dispute over nature, in which different actors are trying to appropriate it culturally and politically in order to determine what kind of eco-social transition should take place and in what direction. In the background of this picture, and once again using the concept of "deep history", we can see the fear of the agricultural sector of losing its "place in the world", understanding that its continuity is threatened in a territory where, as they point out, they and their ancestors have worked for centuries.

The fifth chapter describes the historical evolution and current situation of the second element of pressure on the lagoon: mining, which has developed in the area for centuries. Here, Baños and Baños-González show us the origins of the various discharges into the bay (from the transport of solids by air to the various collapses of ponds containing polluted water, including the more everyday spills that reach the Mar Menor through the area's ramblas), but above all, they offer us a list of measures – some of which are already being implemented – that will make it possible to reduce these spills in the future, thus putting an end not only to the heavy metals entering the Mar Menor, but also to the various health problems suffered by the local population as a result of the mismanagement of mining waste.

With the sixth chapter, signed by María José Alcaraz León, we begin the second part of the book, which we have called Representation, and in which we try to discuss aspects related to the many forms of representation of the Mar Menor that we have at our disposal, which we summarise in three: artistic representation, political representation and legal representation. Alcaraz's chapter acts as a hinge between the two, precisely because it allows us to introduce a new element without leaving the territory: what if we start thinking about the Mar Menor not only as a physical space, not only as an ecosystem, but also as a cultural landscape? And if we do that, what do we have to start "seeing", what do we have to pay attention to? First of all, in order to understand that the cultural landscape is the "product" of the relationship between an ecosystem and the populations that inhabit it, we must take into account not only its ecological values but also its cultural ones. Thus, according to Alcaraz, a cultural landscape has a series of aesthetic values that are the result of this interaction and therefore always depend on the fragile balance that is established between the two. In the case of the Mar Menor, the author argues that it is a cultural landscape to which its inhabitants have developed deep ties, and that the current crises have caused them to feel a profound sense of loss.

The seventh chapter, written by María Ptqk and previously published in Spanish in the book "Reset Mar Menor: Laboratorio de imaginarios para un paisaje en crisis" (Reset Mar Menor: Laboratory of Imaginaries for a Landscape in Crisis), starts from this moment of loss to tell us about an artistic research experience that, from a contemporary perspective, proposed to produce new narratives around the Mar Menor, based on the idea of the "monster" or fabulous creature, on the one hand, and on the introduction of the "monster" or fabulous creature, on the other, on the other hand, the introduction of classical "objectification techniques" – such as cartography or documentary – which, through a critical use, not only questioned the supposed "objectivity" produced, but also tentatively and provisionally illuminated experiences that had previously been left out of the official representations of the conflict. The use of contemporary artistic practices, hybridised with scientific and cultural traditions such as the amateur study of nature (turned into citizen science by the current academic bureaucracy), and a defence of the imagination as a political act combine in the effort to find a way to represent a matter of concern as comprehensively as possible. We do not use the author's terms here, but those used in this introduction, but we believe that we do not betray the spirit of the chapter by framing it in this way. If matters of fact are the result of the historical development of a series of techniques of objectification and representation that have allowed us to visualise them as a fact, that is, as something given (Daston & Galison, 2007), María Ptqk's chapter shows us the effort required to achieve this representation of matters of concern, which has yet to be achieved.

In the eighth chapter, we leave the world of artistic representation to focus on politics. The question we ask ourselves is: how can the Mar Menor be represented politically? Are there institutions capable of giving a voice to natural entities? In this chapter, the authors explore the potential of the Citizens' Assembly as a political space in which such representation might be possible. Drawing on previous experiences, such as the abortion assembly in Ireland or the various climate assemblies held internationally, Gutiérrez, Ballester and Cabello present the results of the experience carried out in our course, where a regional citizens' assembly was simulated to respond to the mandate "For a less eutrophic Mar Menor, what changes are necessary?". As a novelty, the following role-play involved not only human actors representing the different sectors involved in the controversy, but also non-human actors such as natural species (seahorse, broccoli) or the ecosystem itself: the Mar Menor. The conclusions of this exercise do not deny the potential of this method to achieve our objective, but they do point out that there is still a lot of work to be done.

The following two chapters introduce us to the world of the legal representation of nature and, more specifically, the Mar Menor. The first, written by Teresa Vicente and Eduardo Salazar, explains the process and consequences of one of the great achievements of the European environmental movement in recent years: the declaration of the Mar Menor as a legal entity. On the one hand, this signals the beginning of a rupture in our cultural conceptions of the relationship between culture and nature, a crisis in the "cosmology of modernity" – Vicente points out, quoting Latour –, which places us in a new moment in which it is necessary to review the division we have internalised; on the other hand, the chapter announces the emergence of a new generation of subjective rights, the rights of nature, which in turn imply a new paradigm of justice: ecological justice. Law 19/2022, approved by the Spanish Parliament in September 2022, proposed by a popular legislative initiative that collected more than 600,000 signatures in the midst of a pandemic, is the first example of this paradigm shift in Europe, as it is the first time in European legislation that an ecosystem is recognised as a subject of rights.

The second chapter, and the last in this block, takes us to another context: European legislation on the rights of nature and the efforts of some Member States to mitigate its consequences. The case of the Mar Menor is a clear example of how European legislation can be ineffective in protecting an ecosystem. Despite having multiple legal protections at international, European and national levels, it is on the verge of collapse. The European Union is currently debating a new law to restore degraded ecosystems such as the Mar Menor, but at the same time, member state governments are pushing to relax certain rules of the Common Agricultural Policy, which could complicate recovery efforts. In this article, María Giménez analyses the tension between the need to restore the Mar Menor and the pressure to maintain certain economic activities, such as agriculture, that have contributed to its deterioration.

The following chapters are included in the last part of the text, entitled "Inhabit", and they move away from the question of representation to focus on another equally or more pressing issue: the question of the habitability of the territory. In other words, can we imagine a future in which the Mar Menor/Campo de Cartagena ecosystem is a more habitable territory for all those who live there? This leads us to many topics that, at first glance, seem disparate but are all linked to this central question.

Miguel Mesa del Castillo, for example, poses a fundamental question: what is architecture that does not destroy the environment but integrates with it and contributes to its conservation/restoration? In other words, is it possible for architecture to go beyond colonisation and resource extraction? To answer this question, Mesa del Castillo examines the design of La Manga by Antonio Bonet Castellana – an important figure in modern Spanish architecture and a student of Le Corbusier – for the developer Tomás Maestre in the 1950s. Considered an example of "quality" architecture that respects the environment and proposes an occupation of the land that allows the preservation of its natural spaces, Mesa del Castillo argues that this interpretation is the result of a later idealisation of this first design in the light of the actual urban development of La Manga and shows that Bonet's work already contained the seeds of this evolution. In order for this new relationship between architecture and the environment to be possible, the author concludes, a profound reflection is needed that will allow us to distance ourselves from the colonial past, not only in terms of architecture, but also in terms of urban planning and ecology itself.

In her chapter, Mijo Miquel carries out a comparative analysis of three projects focusing on the Albufera of Valencia and the Mar Menor. These projects seek to understand the complex relationships between society and the environment, and how environmental problems are perceived and experienced in these areas. Through different methodologies such as workshops, interviews and the creation of visual narratives, different perspectives and emotions related to the environmental crisis have been identified. However, these projects also reveal the difficulties of constructing common narratives about nature in polarised social contexts, where different visions of the environment generate conflict and tension, and where perspectives on the future and what we consider to be a "habitable environment" are hardly compatible.

In the following chapter, written by Miguel Ángel Martínez, we analyse in depth one of the projects presented in the previous chapter, specifically the first edition of Presentes Densos, which took place around the Albufera de Valencia between September 2020 and June 2021. The text analyses this cycle of conferences in order to reflect on the narratives and practices that come into play in situations of climate emergency and eco-social transition. Martínez stresses the importance of reconnecting with the "common causes" that unite us as a society and allow us to face current challenges, rather than seeking individual solutions or solutions based solely on scientific knowledge. The author argues that while we are ill-equipped to respond to this crisis, we can develop new capacities through collaboration and collective participation. This is what Dense Presents is all about: an attempt to develop new capacities through collective inquiry in the context of an eco-social crisis. This chapter offers us a picture of what those future processes might look like that could be set in motion to learn to live on a damaged planet.

In his chapter, David Soto Carrasco analyses the crisis in the Mar Menor as an example of the consequences of a capitalist economic model that prioritises the accumulation and exploitation of natural resources without considering the long-term environmental and social impacts, using concepts drawn mainly from Jason W. Moore, Kohei Saito and Mark Fisher. Soto shows how practices such as over-farming and cost externalisation have led to the degradation of ecosystems and the loss of quality of life for local communities. However, the author goes a step further in this analysis by showing how these processes, applied to the Campo de Cartagena area, have turned it into a "sacrifice zone" (Juskus, 2023) – a territory where the negative impacts are concentrated in certain regions and populations – affecting their physical, mental and emotional health – which will eventually make it uninhabitable. This chapter concludes by stating that the situation of the Mar Menor reflects a profound environmental injustice that requires a rethink of the economic and social model.

The final chapter of the book, written by Juan Manuel Zaragoza, focuses on the impact of the eco-social crisis on the population living on the shores of the Mar Menor. More specifically, the study focuses on the citizens involved in the defence of the lagoon. The aim is to understand how their emotional well-being is affected by living in an area that has been labelled a "sacrifice zone" (to use Soto's term) and that, according to some discourses, is condemned to suffer a complete deterioration of its living conditions. Using methods derived from participatory artistic research (Nunn, 2022), we first explore their sense of belonging to the territory and, subsequently, how their emotional well-being has been affected by the environmental crisis of 2019 and their response to it. This response reveals an interesting paradox, because while the imaginary projected into the future is that of a recovery of the Mar Menor "as it was before", the political response is structured by a radically innovative measure that actually makes this "as it was before" impossible: the ILP, which has requested legal personality for the Mar Menor. From this contradiction, they manage to escape from nostalgia and offer an imaginative response that implies the creation of a new political community around the Mar Menor as a subject of rights.

As editors of this book, we understand that this diversity of approaches provides a unique overview of the process of ecological degradation of the Mar Menor, but above all, we believe that it contributes to its emergence as a matter of concern, as an event that affects us and forces us to offer a response. The intrusion of the Mar Menor into our lives, as Stengers points out in relation to Gaia (Stengers, 2015), cannot go unanswered.

With this book, we have tried to respond to this intrusion from our commitment to the territory and our obligation as teachers and researchers from the public university – which is just another way of being a citizen, among others. We hope that our response will contribute, at least a little, to understanding what has happened in this region of south-eastern Spain and to countering its consequences. We also hope that it will serve to shed light on similar processes taking place throughout Europe and the world. In the end, as Latour reiterates, we all belong to a single ecological class, and the struggle we must wage for the future is the struggle for the habitability of the planet (Latour & Schultz, 2022). We believe that there are traces in the history of the Mar Menor that allow us to imagine a hopeful end to this struggle. We hope so. The stakes are high.

Notes

1 Anoxic conditions occur when the rate of oxygen consumption exceeds the rate of oxygen replenishment in a water body. This can be caused by factors such as eutrophication, which leads to algal blooms and subsequent oxygen depletion during decomposition.
2 The Spanish Desamortization was a series of laws enacted in Spain in the 19th century that resulted in the expropriation and sale of land and property owned by the Catholic Church, monasteries and municipalities. These lands, known as "manos muertas" (dead hands), were not subject to taxation and were considered unproductive. The government's aim was to increase state revenue, modernise the economy and create a new class of landowning peasants. This process had a profound effect on Spanish society and its economy.
3 The loss of Spain's last colonies in 1898, known as the Disaster of '98, plunged the country into a deep crisis. This defeat sparked a reflection on the causes of Spain's decline and gave rise to a renewal movement known as "Regeneracionismo". The Regeneracionistas, led by figures such as Joaquín Costa, sought to modernise Spain and overcome the structural problems that had led to the loss of the colonies. Their proposals included political, economic and social reforms to create a stronger and more competitive nation. Although Regeneracionismo had a significant impact on Spanish culture and politics, it failed to radically transform the country.

Reference List

Camprubí, L. (2017). *Los ingenieros de Franco: Ciencia, catolicismo y Guerra Fría en el Estado franquista*. Editorial Crítica.

Daston, L., & Galison, P. (2007). *Objectivity*. Zone Books.

Gramsci, A. (2005). *The Southern Question*. Guernica Editions.

Honneth, A. (1998). Democracy as Reflexive Cooperation: John Dewey and the Theory of Democracy Today. *Political Theory*, *26*(6), 763–783. https://doi.org/10.1177/0090591798026006001.

Juskus, R. (2023). Sacrifice Zones: A Genealogy and Analysis of an Environmental Justice Concept. *Environmental Humanities*, *15*(1), 3–24. https://doi.org/10.1215/22011919-10216129.

Latour, B. (2004a). *Politics of Nature: How to Bring the Sciences into Democracy*. Harvard University Press.

Latour, B. (2004b). Why Has Critique Run Out of Steam? From Matters of Fact to Matters of Concern. *Critical Inquiry*, *30*, 225–248.

Latour, B. (2008). *What Is the Style of Matters of Concern? Two Lectures in Empirical Philosophy*. Van Gorcum.

Latour, B. (2018). Outline of a Parliament of Things. *Ecologie politique*, *56*(1), 47–64.

Latour, B., & Schultz, N. (2022). *Mémo sur la nouvelle classe écologique: Comment faire émerger une classe écologique consciente et fière d'elle-même*. Les Empêcheurs de penser en rond.

Morales Gil, A., Rico, A., & Hernández-Hernández, M. (2005). *El trasvase Tajo-Segura*. https://rua.ua.es/dspace/handle/10045/23025.

Nunn, C. (2022). The Participatory Arts-Based Research Project as an Exceptional Sphere of Belonging. *Qualitative Research*, *22*(2), 251–268. https://doi.org/10.1177/1468794120980971.

Sánchez, E. (2019, octubre 20). La asfixia del mar Menor. *El País*. https://elpais.com/sociedad/2019/10/18/actualidad/1571415252_866797.html.

Stengers, I. (2010). *Cosmopolitics I* (R. Bononno, Trad.). University of Minnesota Press.

Stengers, I. (2015). *In Catastrophic Times: Resisting the Coming Barbarism*. Open Humanites Press.

1 The Mar Menor – the transition of an ecosystem over time (its crises in the last 200 years)

Julio Mas Hernández

Introduction

The Mar Menor is a saltwater lagoon located on the south-eastern Mediterranean coast in Spain, in the region known as Campo de Cartagena. Its origins date back around 18,000 years to the Pleistocene epoch when a large bay covered the limits of the eastern part of the Baetic Mountain range, to the north of Escalona, Columbares, Villares, Puerto and Carrascoy mountain ranges, and to the south the Fausilla, Gorda, la Muela and el Algarrobo ranges (see Figure 0.1). Over time, the climate and epicontinental erosion jointly contributed to the gradual clogging of the basin and the formation of rocky barriers, actually fossil beaches, resulting from several transgressions and regressions of ocean levels in different geological eras. The building up of sand on those "barriers" led to the formation of strip of land known as La Manga, isolating the coastal bay from the adjacent Mediterranean Sea, and consequently forming the Mar Menor as we know it today. The weather in the region, the confined water body and its interaction with the continent led to this peculiar ecosystem: transparent waters and severe environmental conditions, with high salinity, high summer temperatures and very low-temperature variations in winter periods.

As is the case of other coastal lagoons, the Mar Menor is triangular in shape. It is home to five volcanic islands which divide it in two sub-basins – north and south. My proposal here is to take a trip down history from the origins of the Mar Menor at the end of the Pleistocene, to the start of the interglacial period when the "ecological process" lasting through to the present day took place, before ending with some of the most obvious threats to it in the Anthropocene.

Prehistoric period

It was not by chance that this coastal lagoon area was colonised by Neanderthals, Hominidae and our other ancestors from very early times. Ecosystems of this type, sheltered from the open sea, with their peripheral wetlands or marshes were rich in game and fish, while also providing a source of water and other essential natural resources to feed the inhabitants. Indeed, two of the most important archaeological sites, namely La Sima de Las Palomas (Torre Pacheco) and Cueva de la Victoria (Cabezo de San Ginés, Cartagena) are, along with El Sidrón (Asturias), Atapuerca

DOI: 10.4324/9781003489078-2

(Burgos) and the Guadix-Baza area (Man of Orce), some of the sites where the oldest Hominidae remains have been found on the Iberian Peninsula.

Some specific examples of the prehistoric epoch can be found in the work by Agustí Ballester et al. (1986) in Cabezo Gordo and the Cueva de Los Mejillones dig near to Cobaticas (Los Belones), dating back to the Upper Palaeolithic Magdalenian cultures. This period ties in with other habitats in the region, such as those in Calblanque and Las Amoladeras (García del Toro, 1986), which were first used by the post-Palaeolithic hunters and fishermen providing them with a diversity of resources in the varied habitats arising around the lagoon.

Relevant background between the prehistoric period and the end of the 19th century

There is abundant bibliography on the terrestrial environment – from the early prehistoric settlements mentioned previously, to the environmental aspects described by Pliny the Elder in his *Natural History* in the year 77 AD, or the flora and fauna, also discussed in "Libro de la Montería" by King Alphonse XI from 1340 to 1350. There are many other references such as Guirao (1859) on the birds in part of Murcia province. However, there are much fewer texts referring to the marine environment there, other than sailing or archaeological records. The oldest cartography of the area is highly inaccurate. Indeed, the map of the "Bishopric and Kingdom of Murcia", dated 1786 shown in Figure 1.1, shows the Mar Menor as a water body that is very isolated from the interior of the continent.

There are, however, references of a certain age referring to fishing, which provide information about the methods used and main species that were caught. In general, lagoon fishing was adapted to the peculiarities of the ecosystem. In the specific case of small confined seas, with channels leading to the adjacent open sea, specific fishing systems have been developed in many parts of the world to take advantage of fish migrations in one way or another, the exploitable populations, trapping them using different specific methods in labyrinths, trapping the catches in a semi-closed area. In the case of the Mar Menor, it is known as "Encañizada", since the labyrinth placed in the so-called "golas" (channels connecting both seas) is built using nets and canes, hence the name in Spanish, comprising the final parts of the fishing system, leading to the codend, which in this region is known as the "corral".

This fishing method has historically been a source of conflict, with lawsuits on production and commercial exploitation rights being fairly common, at least from the times of Alphonse XI, ending in 1910 with the passing of the "Fishing and use of fishing techniques Regulation in the Mar Menor". The 18th-century map as shown in Figure 1.2 shows the location of the "golas" and a jurisdictional division of the waters of the Mar Menor: to the left belonging to the city of Murcia and to the right belonging to Cartagena. Likewise, the watchtowers and forts are shown which defended the sparsely populated coastline from different attacks throughout history by the Punics, Vikings and later the Berbers. In the "Historical Dictionary of National Fishing Arts" (1773), Sáñez Reguart also refers to the "encañizadas",

Figure 1.1 Map of the Bishopric and Kingdom of Murcia divided into its districts. Based on the map printed by Felipe Vidal y Pinilla and the private recollections of individuals. By the geographer D. Thomas López, pensioner of the S.M. and the Royal Academy of San Fernando. 1768.

whereas, much later in the "Dictionary of the Art of Fishing in Spain and Its Possessions" (Rodríguez Santamaría, 1923), there were many texts, drawings and figures referring to the different arts of fishing and the main fish species that were caught.

The environmental impacts that could be considered relevant within that long period are mainly due to human exploitation. The one that had the most impact during that period was undoubtedly the Sierra de Cartagena-La Union mine, which is discussed in greater detail in Chapter 5. Exploitation began at the same time as the first settlers inhabited the area in protohistorical times, i.e. in the iron and bronze ages, and such activity continued through to the second half of the 20th century.

Another age-old activity in the region is salt mining. There are many salt evaporation ponds that prove that small-scale salt making was carried out by the first

Figure 1.2 Plan for the distribution of fishing between Murcia and Cartagena. 18th century. Archivo de la Real Chancillería de Granada/059CDFI//MPD n.º 18.

settlers with techniques being perfected later during the period of the Roman Empire. The small bays around the inner part of the Mar Menor were isolated from the lagoon to speed up evaporation. That is the case of the Patnia evaporation pond (located on the northern edge of the lagoon), which Alphonse X incorporated to the Spanish crown in 1266.

In short, apart from the impact of mining and the pollution associated with it, neither the morphology nor the environmental conditions of the lagoon were seriously affected in environmental terms throughout that long period. It is obvious that stability was not constant though, and as is the case of the rest of the region, it was subjected to deforestation, gradual population growth and arable and livestock farming, in addition to some severe weather phenomena such as the "Cold Drop" (high impact rainfall events) or "DANAS" (isolated high altitude depressions) causing floods and the subsequent washing down of inland sediments that were deposited in the lagoon.

First period of environmental crisis (late 19th century and early 20th century)

This period spans the final years of the 19th century through to the middle of the 20th century, and we shall see how the aforementioned conditions had not undergone any major changes since the Middle Age. Two significant morphological changes around the inner perimeter did take place at the start of the period, with the artificial increase in the "mota" (wall) separating two salt mills in San Pedro del Pinatar, to the north (Figure 1.3) and the start of another salt works on a small golf course to the south of Marchamalo (Figure 1.4). Influence by human activity gradually increased at a faster rate after that period.

Some new settlements were also established during the same period, such as Santiago de la Ribera, where much of Murcia's upper class built their holiday homes (Alonso Navarro, 1989), and the first plans for economic and social development of the area were defined. Indeed, José María Barnuevo, Senator and Deputy of the Courts made a speech in Congress on May 21st, 1897 describing the deficient conditions of the Mar Menor's health and the poor living conditions of the inhabitants there, mainly subsistence farmers, a constant shortage of fish catches, a lack of wastewater treatment infrastructures and frequent outbreaks of malaria due to the numerous marshes around the shores. He put forward opening it up to the Mediterranean Sea, building a fishing port and improving communications to connect the coastal villages with the larger urban areas where there were better social and economic services (Barnuevo, 1898).

Heavy metals from mining in the mountains were still draining into the southern part of the Mar Menor, which had intensified in the later years of the 19th century and early 20th century. Those heavy metals are still found in the sediment and biota in the lagoon (see Chapter 5).

Second period of environmental crisis (mid-20th century to the early 21st century)

Three factors mainly affect this second period: the opening of "El Estacio" channel; generalised urban development throughout the region, both along the inland part of the coast and on the strip of land known as the Manga; and the change from traditional unirrigated arable farming to intensive irrigated arable farming. The latter underwent a major increase with the arrival of irrigation water from the River Tagus-Segura water transfer system which was opened in 1979.

From an oceanographic point of view, the parameters that best characterised the lagoon over time were the temperature and salinity, which in turn conditioned the flora and fauna and the communities inhabiting the shores. Another differentiating element was its isolation and the variations in the amounts of water exchanged between the Mar Menor and the Mediterranean Sea.

In this sense, there are initially two clearly different stages: the first spans the early years of the 20th century through 1970 with severe isolation and salinity, which occasionally exceeded 50 PSU (practical salinity units) and a minimum water temperature of 6°C. The second stage began with the opening of the Estacio channel (1974–1975), leading to the main ecosystem changes before the current crisis, which led to a fall in the salinity level to 44 PSU and a minimum temperature of 9°C (Arevalo, 1988).[1]

The opening of the Estacio channel marked the change between both stages. It consisted of dredging and extending the existing channel to allow access to the Mar

Figure 1.3 Aerial photograph of the north of the Mar Menor, corresponding to the salt marshes of San Pedro del Pinatar.

Figure 1.4 Aerial photograph of the south of the Mar Menor, in the area corresponding to the salt marshes of Marchamalo.

Menor by vessels with deeper drafts. The extension of the channel meant that there was a greater exchange of water between the Mar Menor and the Mediterranean, which explains the fall in salinity, and as we shall see later on, the arrival of new species that were unable to withstand the previous high salinity level, but which were now able to thrive.

It was also in this period when the first truly thorough oceanographic data were compiled, thanks to the work of two Spanish oceanographic researchers, namely Antonio Arévalo Arocena and Jesús Aravio Torres, who focussed their studies on the salinity levels and the temperature of the water in the Mar Menor (Arévalo & Aravio-Torre, 1969; Aravio-Torre & Arévalo, 1975). In addition to the aforementioned papers, the work on the ichthyofauna in the Mar Menor is worthy of mention, from the early pioneering surveys by Francisco de Paula Navarro (Navarro, 1927), Luis Lozano Rey (1947a, 1947b, 1952) and Fernando Lozano Cabo (1954, 1969, 1979). Those surveys highlighted the predominance of Mugilidae (up to six species) and gilt-head bream in the first stage of isolation, whereas, in the second one, the number of species increased and the preponderance of the dominant groups decreased.[2] As a result of the growing interest in the Mar Menor, an Oceanographic Station was founded in 1966 in San Pedro del Pinatar, which was a branch of the Spanish Oceanographic Institute, directed by Joaquín Ros Vicent, and which would later become a research centre of reference on the lagoon (Pérez de Rubín, 2014).

On the other hand, it is worth remembering that massive colonisations took place during that period, with significant demographic changes, different taxonomic groups such as jellyfish (*Cotylorhiza tuberculata* and *Rhizostoma pulmo*), Ctenophora and crustaceans such as blue crabs (*Callinectes sapidus*). At the same time, we saw the practical extinction of the seahorse (*Hippocampus guttulatus*) and the more recent expansion and subsequent almost disappearance of the pen shell (*Pinna nobilis*), of which there was a more drastic antecedent with the appearance of the flat oyster (*Ostrea edulis*). This bivalve mollusc peaked with a population of over one million individuals, before being reduced to negligible numbers.

These alterations of the "normal" cycles of the lagoon have contributed to its imbalance, pushing its mechanisms of resilience, mainly high salinity and water transparency, to the point where they were severely compromised. An example of this was at the end of the 90s with the EUROGEL[3] project, financed by the European Union, in which the presence of over 80 million individual jellyfish was documented, which in subsequent years fell to negligible figures.

As for the vegetation in the lagoon, during the first "isolation" stage, there was fairly little vegetation, with only some marine phanerogam such as the *Cymodocea nodosa* and *Ruppia cirrhosa*, as well as some other green algae. Following the opening of the Estacio channel, other plants colonised the Mar Menor, some macrophytes such as *Caulerpa prolífera*, which eventually practically covered the entire bed of the Mar Menor. In the specific case of this species, its contribution to the silting up and increase in the content of organic matter in the sediments and displacement of the previous communities was fundamental to understand certain episodes of anoxia and other subsequent processes.

This period could be considered one of semi-stability, or better said, one of adaptation to the simultaneous effects of the three initial factors mentioned previously, which explain the considerable variability of the Mar Menor's environmental balance.

Third period of environmental crisis (early 21st century to 2023)

It was not until the turn of the 21st century that a more turbulent period began. As was true in the previous period, we can identify two stages. One started around 2000 and ended in 2016, whereas the second, most critical stage, began in 2016 and is still ongoing. It is during this period that the eutrophication process of the aquatic ecosystem intensified, with increases mainly from farming (85%), enhanced by the high temperatures in Spring 2016, leading to a multiplication by over 100 of the chlorophyll compared to the values recorded in the last 20 years. This massive blooming of phytoplankton, accompanied by other micro and nano organisms, such as fungi and viruses, leads to the so-called "pea soup", marking the start of the second period. The entire water column turned green, preventing light from passing through the water and therefore the plants on the bottom are unable to photosynthesise. This effect accounted for a loss of 85% of the marine prairies associated with the benthal macrophytes covering the 135 km^2 of the lagoon's bed.

Following on from that, the build-up of decomposing organic matter on the bottom gave rise to an unbridled demand for dissolved oxygen in the water column until it was completely exhausted, thereby leading to generalised anoxia in the lagoon. This worsening of the situation meant the water became cloudier, silt and algae appeared on the shores, which evolved from stages of hypoxia through to anoxia and eventually to euxinia (sulphidic water), ending in mass death of fish, crustaceans, molluscs, etc., between 15th and 16th of August 2019, with this event featuring on international news.

It is also important to highlight the DANA also known as the cold drop on October 12, 2019, one of the heaviest in recent decades, when it rained heavily over the open sea raising the water level of the Mediterranean Sea thereby preventing drainage of excess water from the Mar Menor to quickly discharge its excess water and the consequent flooding of the inland coastal areas.

More local episodes of anoxia have also taken place in the southern part of the Mar Menor, such as around Isla del Ciervo, with less severe mortality of flora and fauna between 2021 and 2022. The problems of massive algae blooms were particularly visible in 2023 along some parts of the coast and the silting of the beaches and coasts.

Current situation and discussion on future evolution

In the last decade, 2013–2023, the two main features that explain the unique conditions of the Mar Menor, i.e. salinity and temperature, have undergone enormous variations – salinity, due to the higher phreatic level of the aquifer under the Campo de Cartagena region, bordering the lagoon. The extra supply of water, resulting from the change from traditional unirrigated farming to intensive irrigated farming, means that water from the aquifer enters the Mar Menor and modifies its salinity. In turn, the temperature range of the water column has also undergone significant variations owing to the specific conditions of the region, but also most likely due to the effects associated with climate change.

In addition to this, there have been some very dry years in the study period and particularly strong heatwaves. The 2024 cycle is considered to be the hottest since records began. These heatwaves have been repeated around the planet, and in the Mediterranean region, they have especially affected the mid sector. Despite the said conditions, it can be said that the generalised periods of anoxia have not been repeated in the lagoon in the last two years (2022–2023), although the effects of environmental crisis have left their mark, with an alarming decrease in fish catches, rarefaction or loss of some more vulnerable species populations, such as seahorses or pen shells, and in general a notable deterioration of the ecosystem as a whole, both in the lagoon itself and on the adjacent shores and banks, including the marshes.

Parallel to this, new problems are appearing that also affect the balance of the ecosystem, such as new invasive species, problems associated with emerging pollutants (hormones, chemicals or other substances that are not treated in current wastewater treatment systems), and also a generalised problem in the ocean:

Figure 1.5 A white patch originating at the mouth of the Rambla del Albujón and now extending beyond Isla Perdiguera and El Barón. It was first noticed in the early months of 2023, although there were signs of anomalous colouring before then. Its origin and the identification of its causes are still under investigation.

breakdown of plastics to form microplastics, which on entering the trophic chain are eventually consumed by humans and in certain species the levels of heavy metals exceed legal limits. Although only occasionally, some Massive Algae Formations have taken place, some of them being potentially hazardous due to the presence of poisonous dinoflagellates.

The latest threat nowadays is caused by the complex interaction related to fluctuations in the phreatic levels, the quality of the groundwater, dissolution of materials at different geological levels with particularly high concentrations of nitrates, and the consequent influence of them on the Mar Menor water column (this is also closely linked to intrusion of seawater in the continent). This mainly affects the salinity and pH variations and the huge "white stain" (Figure 1.5) arising at the mouth of the El Albujón ravine expanding southwards causing a shadow effect, similar to the "pea soup" and preventing light from reaching the bed, consequently causing the death of the light-dependent benthic species. Another consequence, relatively unknown or at least not clearly identified, is a disruptive phenomenon in the primary production trophic chain, which could perhaps alter the coloration of the water bodies, as is the case of "red tides" when massive blooming of poisonous phytoplankton take place, in this case, mainly dinoflagellates.

To end this chapter, we are going to consider some thoughts about the future, mainly dealing with two particularly relevant aspects. One of them is the clogging/eutrophication of the lagoon, where transformation of uses and territorial planning are urgently required to correct the errors of the past.

In this context, some recent novel research projects, data collection and information contained in the Priority Actions Framework for recovery of the Mar Menor (MITECO, 2024) by MITECO, plus the IEO/CSIC monitoring reports in different multidisciplinary lines of action (IEO, 2024) provide some very relevant, up-to-date information. In its last report in February regarding the status of the water column, it highlights that 2023 was an apparently stable period, with the exception of the area covered by the white stain. This stability cannot be construed as recovery, since it could just be a transitional stage caused by climate factors, for example. On the other hand, the increasing pH trend suggests that the ecosystem has not stabilised and the feedback from biological monitoring activities does not show a recovery of marine plant habitats or species on the sea bed. Parallel to this, MITECO announced a budget increase to carry out the planned actions, increasing it from 484 million Euros to 675 million to be implemented between now and 2027. Among the new actions planned for that period are the restoration of the Sierra Minera, reforesting and environmental restoration of the high and mid basin of the Campo de Cartagena region, new calls for grants to adapt the municipal drainage networks with separation of rainwater, and a line of subsidies to adapt the livestock sector, in addition to those already allocated to the arable sector.

The other critical point is the physical continuity and maintenance along the Mediterranean coastline, threatened by rising sea levels caused by global climate change on the planet. Apart from the collateral effects of heatwaves identified in those forecasts, increased aridity and low levels of precipitation are also expected to play a role.

In a very recent report by the NASA Sea Level Change Team (2024), the coast in the Region of Murcia will be the third space in Mediterranean Spain where the sea level will rise the most, just behind Barcelona and Valencia. The estimates made by NASA'S scientists suggest that if nothing is done to reduce CO_2 emissions, and in a scenario in which the rise in temperature will be of up to 5 degrees, the sea level in Cartagena will be 12 centimetres (cm) higher than its current level by 2030, which will double by the mid-century to reach 27 cm and 79 cm by 2100. These forecasts and scenarios are particularly concerning with a view to maintaining the dune system forming La Manga, and therefore isolation of the Mar Menor from the Mediterranean Sea.

Heatwaves also have a collateral effect on other aspects that will affect the oceanographic parameters of the Mediterranean Sea, and therefore the Mar Menor too. Based on available information, the European Environment Agency (Vecchio et al., 2023) warns that the loss of dissolved oxygen in water is increasing: 25% of monitored marine areas in Europe have low concentrations, less than 6 mg/l, according to data compiled between 2011 and 2022, i.e. "under the minimum threshold to support life under stress". As for Spanish water, the levels are under 2 mg/l in certain zones. This threshold is considered to be hypoxia, with the data collected from around the Balearic Islands and the coast of Alicante being particularly concerning, as they are closely related to the water in the Region of Murcia. Oceans have lost 2% of their oxygen content since 1950, which is the result of warming marine water caused by climate change, and the effluent containing nutrients from farming, aquaculture and wastewater. To end with, I shall emphasise that all the aforementioned items are worse in arid regions or those undergoing desertification, as is the case of the southeast region of the Iberian Peninsula, or in relatively small confined water bodies, as is the case of the Mar Menor.

The significant increase in the number of monitoring studies and research projects on the Mar Menor in recent years has provided abundant information and new data that we hope will allow long-term solutions to be implemented. Some more restrictive legislative measures have also been implemented and areas protected from intensive farming have been established in the regions bordering the lagoon, and this has undoubtedly at least contained the constant deterioration process that was observed during the final stage of the third crisis period. The decisive action by MITECO based on nature, on the other hand, has helped to identify satisfactory responses to the problems that have been affecting the territory for many years, such as the ravines transporting effluent from the Sierra Minera to the Mar Menor. In turn, the Regional Authorities are defining significant programmes to breed endemic species of the Mar Menor in captivity, whereas organised civil society is undertaking territory protection and recovery work, such as recovery of the Marchamalo evaporation ponds by the ecological association ANSE.

All the data points to the possibility of a more promising future for the Mar Menor, with an ecosystem that will recover its balance and strengthen its resilience mechanisms. This, however, is not enough. The threats to the Mar Menor, particularly originating through global climate change, require much quicker, more decisive action to tackle the problems before they reach the point of no return.

Notes

1 As mentioned earlier, salinity is a fundamental factor that explains many of the specific features of the Mar Menor, and therefore decreases such as the one described here or the one that took place in Spring 2022 when an average of 39.6 PSU was recorded on the surface and 39.7 PSU on the bottom, similar to the levels in the Mediterranean Sea, are monitored very closely. The 2019 DANA caused a considerable decrease in the mean salinity of the lagoon, and the previous mean values (over 44 PSU) did not recover until 2023 (Ruiz Fernández, 2023, p. 14).

2 Many other surveys and studies were conducted during the same period on the different biological communities in the Mar Menor (phytoplankton and zooplankton, zoobenthos and other taxonomic groups), as well as on other environmental factors affecting the water body, such as the climate and modifications to certain geographical and topographical elements. All the aforementioned data, and the evolution of the parameters, can be found in Mas (1996).

3 European gelatinous zooplankton: mechanisms behind jellyfish blooms and their ecological and socio-economical effects (EUROGEL), EVK3-CT-2002-00074.

Bibliography

Agustí Ballester, J., Axbiol Noria, S., Gilbert i Clos, J., Martínez Navarro, B., Menéndez Cabrera, E., Moyá-Sola, S. (1986). Paleontología del Sureste de la Península Ibérica. *Historia de Cartagena* (Tomo I, pp. 129–164). Mediterráneo.

Alonso Navarro, S. (1989). *Santiago de la Ribera: cien años de historia, 1888-1988*. Ayuntamiento de San Javier.
Aravio-Torre, J., Arévalo, A. (1975). La salinidad del Mar Menor, sus variaciones. Algunas consideraciones sobre el intercambio de aguas con el Mediterráneo. *Bol. Inst. Esp. Oceanogr.*, (146), 3–20.
Arévalo, A., Aravio-Torre, J. (1969). La salinidad de las lagunas litorales. El Mar Menor (Murcia). *Bol. Inst. Esp. Oceanogr.*, (139), 1–27.
Arevalo, L. (1988). El Mar Menor como sistema forzado por el Mediterráneo. Control hidráulico y agentes fuerza. *Bol. Inst. Esp. Oceanogr.*, 5(1), 63–96.
Barnuevo, J. (1898). *El Mar Menor. Breves consideraciones sobre su estado actual y su porvenir*. Imprent. y Litigr. del Depós. de la Guerra.
de Rubín, J. Pérez, Santamaría, J. (2014). *100 años investigando el mar. El Instituto Español de Oceanografía en su centenario (1914 – 2014)*. IEO.
García del Toro, J.R. (1986). Los cazadores-pescadores postpaleolíticos. Sus asentamientos hasta el Eneolítico final. *Historia de Cartagena* (Tome II, pp. 163–174). Mediterráneo.
Guirao, A. (1859). Catálogo metódico de las aves observadas en una gran parte de la provincia de Murcia. *Bol. Re. Acad.Cienc.* (Tome IV, pp. 511–560).
Lozano Cabo, F. (1954). Notas sobre una campaña de prospección pesquera en el Mar Menor (Murcia). *Bol. Inst. Esp. Oceanogr.*, (66), 1–34.
Lozano Cabo, F. (1969). La Fauna ictiológica del Mar Menor. Generalidades y claves de determinación de las especies. *Bol. Inst. Esp. Oceanogr.*, (138), 3–47.
Lozano Cabo, F. (1979). *Ictiología del Mar Menor (Murcia). Los Fisóstomos*. Universidad de Murcia.
IEO (2024). *Mar menor: monitorización e investigación*. Instituto Español de Oceanografía. https://www.ieo.es/es/mar-menor.
Lozano Rey, L. (1947a). Peces Ganoideos y Fisóstomos. *Mem. R. Acad. Cienc. Exact. Fis. Nat.*, (11), 1–839.
Lozano Rey, L. (1947b). Peces Fisoclistos. Tercera Parte, Subserie torácicos. *Mem. R. Acad. Cienc. Exact. Fis. Nat.*, (14), 1–613.
Lozano Rey, L. (1952). *Los peces fluviales de España*. Ministry of Agriculture Madrid.
Mas, J. (1996). El Mar Menor. Relaciones, diferencias, y afinidades entre la laguna costera y el Mar Mediterráneo adyacente. *Micro. Fich. Inst. Esp. Oceanogr.*, (7), 1–301.
MITECO (2024). *Marco de Actuaciones Prioritarias para la recuperación del Mar Menor (MAPMM)*. https://www.miteco.gob.es/es/ministerio/planes-estrategias/mar-menor/marco-actuaciones-prioritarias.html.
NASA (2024). *Sea Level Projection Tool-NASA Sea Level Change Portal*. https://sealevel.nasa.gov/ipcc-ar6-sea-level-projection-tool.
Navarro, F.d.P. (1927). Observaciones sobre el Mar Menor (Murcia). *Not. y Res. Inst. Esp. Oceanogr.*, Serie II, (16), 1–63.
Rodríguez Santamaría, B. (1923). *Diccionario de Artes de Pesca de España y sus Posesiones*. Sucesores de Rivadeneyra.
Ruiz Fernández, J.M. (coord.) (2023). *Proyecto Belich: seguimiento, estudio y modelización del estado del Mar Menor. Informa de actualización de resultados del programa de seguimiento del estado del Mar Menor (octubre 2023)*. IEO.
Sáñez Reguart, A. (1773). *Diccionario histórico de las artes de pesca nacional* (Vol. III). Imprenta de la Viuda de Don Joaquín Ibarra.
Vecchio, A., Anzidei, M., Serpelloni, E. (2023). Sea level rise projections up to 2150 in the northern Mediterranean coasts. *Environmental Research Letters*, *19*(1), 014050. https://doi.org/10.1088/1748-9326/ad127e.

2 Ecosystems and biodiversity in the frontiers of the Mar Menor Lagoon

Birds as alert bioindicators of environmental change

Francisco Robledano, Encarna Guillén, M. Francisca Carreño Fructuoso, Antonio Zamora López and Gustavo A. Ballesteros Pelegrín

Introduction

The Mar Menor (Murcia, SE Spain) is the largest coastal lagoon in the western Mediterranean. The main basin (135 km^2) is a transitional water system heavily stressed by the changes occurring in its watershed which together with it constitute a complex socio-ecological system (Mar Menor Socio-Ecological System, henceforth MMSES). From the 1980s until the first decade of the present century, ecologists have warned about an eutrophication process of mainly agricultural origin that changed its biodiversity and environmental quality (Robledano et al., 2008; Martínez Fernández & Esteve Selma, 2020). From 2015 to 2016 onwards, it experienced a succession of environmental crises including phytoplankton blooms and the loss of large areas of submerged meadows, and even episodes of mass mortality of aquatic fauna (Ruiz-Fernández et al., 2022). On the terrestrial frontiers of the lagoon, a set of wetlands perform an important role in protecting it from the pressures driven by the surrounding land uses (Álvarez-Rogel et al, 2020). There has been a trade-off between the nutrient removal function and the ecological value of these wetlands (Martínez-Fernández et al., 2014). The biodiversity of the MMSES wetlands has an intrinsic value but also plays a key bioindicator role, as an alert system that anticipated the risk of eutrophication of the lagoon several decades before (Robledano et al., 2008, 2011).

The typology of wetlands found around the Mar Menor includes environments located on its terrestrial shoreline (salt pans and other perilittoral wetlands, including those mentioned in the previous paragraph). But in its maritime frontiers, there are also wetlands associated with the communications of the lagoon with the Mediterranean (inlets or "golas"). These latter have been historically managed as traditional fishing devices (Encañizadas) compatible with the preservation of their biodiversity. Since the 1970s, however, most were dredged and artificialized but the few surviving are home to highly exclusive habitats, unique in the Mediterranean due to their quasi-tidal functioning. These fragile ecosystems still face several

DOI: 10.4324/9781003489078-3

threats including proposals for tourist infrastructures (Robledano et al., 2018) and dredging as a solution – clearly absurd – to the eutrophication of lagoon waters (García-Oliva et al., 2018; Pérez-Ruzafa et al., 2018). The birds of the MMSES exploit these aquatic habitats for activities such as feeding, roosting or nesting (Robledano et al., 2010; 2011; Ballesteros et al., 2021). Our main objective is to summarize the bioindicator value of wetland avifauna in relation to environmental changes whose main driver has been the agricultural transformation of the watershed, highlighting those taxocenoses that should be prioritized in monitoring schemes.

Study area

The MMSES, located on the eastern coast of the Region of Murcia (SE of Spain), includes the Mar Menor Lagoon, its surrounding wetlands and the catchment that slopes to the lagoon, separated from the Mediterranean by a sandy bar (La Manga), 22 km long and between 100 and 900 m wide, crossed by five shallow channels or golas (Figure 2.1). The continental part of this MMSES is a basin of 1,270 km^2, drained by various ephemeral watercourses partly blurred near the lagoon into which they flow. It is the largest coastal lagoon in the western Mediterranean, including five volcanic islands within a surface area of 135 km^2, with a volume of 610 hm^3, an average depth of 4.5 m and a maximum depth of 7 m (Guaita-García et al., 2021).

The MMSES houses the most complete and extensive representation of wetlands in the Region of Murcia, with four main types: the coastal lagoon itself (the only regional example) and three other surrounding it in different sectors: *crypto-wetlands* or hidden seepages (saline steppes and other phreatophytic formations, mostly without a visible water table), coastal salt pans and quasi-tidal marshes), with an additional surface area of approximately 10 km^2 (Vidal-Abarca et al. 2003).

Methods

After a brief synthesis of the ecological values associated with these wetlands, we will focus on avifauna as one of the main land-based bioindicator taxocenoses and the most popular and longest record of biodiversity. Examples of the response of bird communities to external pressures will be presented, such as the changes observed in the bird populations of the main lagoon water mass, or the decline of some characteristic species of perilittoral *crypto-wetlands* as an indirect effect of agricultural transformation. It should be noted that the ornithological monitoring of these wetlands is far from being continuous, since it has been largely based on citizen science and only temporarily on academic research funded by official programs or specific contracts with the environmental administration. Methods include (see also Table 2.1):

- Boat census along a predefined route inside the lagoon basin, performed yearly (every January) as a part of the International Waterbird Census (IWC). Methods

Table 2.1 Summary of ornithological indicators selected from the main studies carried out by the authors in the MMSES wetlands, with indication of the explanatory factors or variables showing a significant association with waterbird changes. "Main response" is preceded by the sign of the change in each particular indicator (or group of indicators)

No.	*Wetland type and spatial coverage*	*Temporal scale (time span)*	*Sampling method*	*Bird bioindicator*	*Explanatory factor(s) or variable(s)*	*Main response*	*Source*
1	Mar Menor Lagoon (whole basin): 13.500 ha	Long term (1970'–present)	Boat census	Red-breasted Merganser (*Mergus serrator*)	Nutrient load	(−) Long-term decline and virtual disappearance	Robledano et al. (2011)
2	Mar Menor Lagoon (whole basin): 13.500 ha	Long term (1970'–present)	Boat census	Eurasian Coot (*Fulica atra*)	Nutrient load	(+) Peak in 2005; (−) subsequent decline and disappearance	Robledano et al. (2011)
3	Mar Menor Lagoon (whole basin): 13.500 ha	Short term (2007 vs. 2018)	Boat census (short transects)	Summer (Apr–Sep) waterbird counts	Lagoon eutrophic crise (B/A)	(+/−) Various responses	Robledano et al. (2019)
4	Mar Menor Lagoon (whole basin): 13.500 ha	Short term (2006–2008 vs. 2016–2019)	Boat census (short transects)	Winter (Oct–Mar) waterbird counts	Lagoon eutrophic crise (B/A)	(+/−) Various responses	Robledano et al. (2019)
5	Mar Menor Lagoon shoreline (1 km wide band)	Short term (2006–2009 vs. 2016–2019)	Land-based census (500 × 1,000 m units)	Winter (Oct–Mar) waterbird counts	Lagoon eutrophic crise (B/A)	(+/−) Various responses	Farinós-Celdrán et al. (2017) and Robledano et al. (2019)
6	Marina del Carmolí saline steppe: 318.6 ha	Long term (1984–2008)	Transect (40 m wide band on each side)	*Alaudidae*	Saline steppe vs. salt scrub surface	(−) Decline associated to steppe reduction	Robledano et al. (2010)

7	Marina del Carmolí saline steppe: 318.6 ha	Long term (1989–2022)	Transect (40 m wide band on each side)	*Alaudidae*	Water table	(−) Decline associated to phreatic level rise	Guillén et al. (2023)
8	Encañizadas (lagoon inlet): 181 ha	Short term (2014/15 vs. 2018/19)	Land-based census of predefined sectors	All waterbirds	Lagoon eutrophic crise (B/A)	(=) No significant change	Zamora (2015) and Robledano et al. (2019)
9	All peripheral wetlands (active saltpans & other)	Long term (2005–2022)	Census of breeding pairs	Kentish plover (*Charadrius alexandrines)*	Active protected area management (Y/N)	(−) Decrease irrespective of protection status and management	Ballesteros (unpublished)

are described by Hernández-Gil and Robledano-Aymerich (1997), and results are reported and analyzed in Robledano et al. (2011), Martínez-Fernández et al. (2014) and Robledano et al. (2019).

- Monthly census of aquatic birds during short periods before and after the main eutrophic crisis of 2015–2016, performed either inside the lagoon (short boat transects distributed along a main route) or in their littoral zone (1 km belt). Results are reported in Farinós & Robledano, 2010), Farinós-Celdrán et al. (2017) and Robledano et al. (2019)
- Census of wintering waterbirds in 11 predefined sectors of the remaining well-preserved *golas* (Encañizadas del Ventorrillo y La Torre), performed in the winters 2014/2015 and 2018/2019. Methods and results are presented in Zamora (2015) and Robledano et al. (2019).
- Census of breeding waterbirds in all the peripheral wetlands of the MMSES (number of breeding pairs), performed by G. A. Ballesteros (unpublished data).
- Winter and spring census of terrestrial birds (mainly passerines) along line transects with a predefined counting band, expressed as IKA (Kilometric Abundance Index). Methods and results are presented in Robledano et al. (2010), Robledano et al. (2019) and Guillén et al. (2023).

Main wetland habitats and ornithological responses

In addition to its ornithological values, for which the Mar Menor wetlands have been designated as Special Protection Areas (SPA) in application of the Birds Directive 2009/147 EC, they are important at the ecosystem level. As a whole, the Mar Menor Lagoon and its perilagunar wetlands contain 15 types of habitats of community interest, included in Annex I of the Habitats Directive 92/43/EEC, of which the following are prioritary (*):

- 1150* Coastal lagoons (the entire bottom of the lagoon basin and the smaller waterbodies of the peri-lagunar wetlands)
- 1510* Mediterranean saline steppes (mainly in perilagunar crypto-wetlands)

In the Encañizadas (the type of man-made wetland located in the *golas*), we also find habitat 1140, defined as "muddy or sandy plains that are not covered with water when there is low tide". This habitat is unique not only on a regional scale, but also on a national one, since it is only known in very few places on the Spanish coast (Belando et al., 2014).

It is not the objective of this chapter to analyze how the ecological state of these habitats as such has evolved, but as will be explained later, it will highlight some changes that seem relevant for bird communities. An example is the physiognomic modifications of steppe habitats (because of hydrological changes in the entire watershed) that can explain the numerical trends of some wetland birds. With regard specifically to the ornithological values, SPAs for Birds have been designated in the MMSES:

- ES0000175 Salinas y Arenales de San Pedro del Pinatar, classified as Bird SPA in 1999 for meeting the numerical criteria for the breeding species Black-winged

Stilt (*Himantopus himantopus*), Pied Avocet (*Recurvirostra avosetta*), Little Tern (*Sterna albifrons*) and Gull-billed Tern (*Gelochelidon nilotica*). Includes the largest and active saltpans of the MMSES (https://www.iepnb.es/eu/areas-tematicas/espacios-protegidos/ES0000175_LIC-ZEC%20y%20ZEPA/salinas-y-arenales-de-san-pedro-del-pinatar)
- ES0000260 Mar Menor, which includes the lagoon and the main perilittoral wetlands, classified as SPA in 2001 for meeting the numerical criteria for breeding Black-winged Stilt, Little Egret (*Egretta garzetta*) and Short-toed Lark (*Calandrella rufescens*). The latter species is typically associated with salt steppe habitat and other low vegetation cover halophilic environments (https://www.iepnb.es/areas-tematicas/espacios-protegidos/ES0000260_ZEPA/mar-menor)

In addition to these species, the wetland complex is home to a great diversity of birds, playing an important role in the conservation of aquatic species at different geographical scales, standing out the local breeding populations of gulls, terns and waders such as the Kentish Plover (*Charadrius alexandrinus*). Likewise, it represents as a wintering or refuge area for many birds breeding in northern latitudes or inland wetlands, among which can be highlighted the once important populations of *Mergus serrator* (Red-breasted Merganser), the regular presence of *Phoenicopterus ruber* (Greater Flamingo) and the still important numbers of wintering *Podiceps nigricollis* (Black-necked Grebe), among others (see references in Figure 2.1). The main result regarding the monitoring of avifauna is the decrease or disappearance of some of these taxa, whether aquatic or steppe dwelling ones. Table 2.1 displays some representative trends of avian bioindicators in the MMSES wetlands and the explanatory factors or variables (descriptors of eutrophication) showing a significant association with them.

Indicators 1 and 2 show long-term changes in representative waterbird species wintering in the lagoon's water mass. The first one is a diving and piscivorous species (Red-breasted Merganser) that, unlike other waterbirds with similar ecology (Great Cormorant and Podicipedidae), has not been favored by the eutrophication process that has enhanced the productivity of the lagoon. Robledano et al. (2011) showed a negative response of this species to the estimated input of nutrients (agricultural N), changing from a dominance in the waterbird community of the original oligotrophic lagoon, rich in its preferred fish prey, to an almost testimonial presence in winter census (Martínez-Fernández et al., 2014). However, the explanation for its virtual disappearance also seems attributable to other factors. Recent studies point to the possible influence of climate change on the displacement of its wintering areas towards the north and east of Europe, closer to the breeding grounds (Pavón-Jordán, 2017). The tempering of winter conditions would have shifted the wintering areas of many waterbird species to the north (Lehikoinen et al., 2013; Moss, 2015; Cox et al., 2023). Many of these studies are based on community science, which highlights the value of integrating citizen data with academic science in the interpretation of species trends and spatial shifts.

The second species (Coot) is an herbivore whose positive correlation with nutrient loading demonstrates that eutrophication effects are not only manifested in piscivores. The peak abundance of Coot has also been related to a phase of higher contribution of nutrients from urban origin (untreated wastewater), a type of organic pollution that reached the lagoon through its main entrance channel, the Albujón ephemeral watercourse (García-Pintado et al., 2007), converted to a permanent drainage channel of agricultural waters. Coots were mainly restricted to the area of influence of that channel, close to its outlet, in numbers that increased from 1998 to 2006. Whether due to urban wastewater, diffuse or concentrated agricultural pollution, or to all of them, such response has as proximate explanation the proliferation in that area of algal mats that constitute the Coot's main food source. Eutrophication commonly causes a deterioration of submerged macrophyte beds consumed by herbivores (Noordhuis et al., 2002), but in turn can favor opportunistic macroalgae (Krause-Jensen et al., 2008; Pérez-Martín, 2023), a source of food for generalist phytophagous species like Eurasian Coot (Perrow et al., 1997; Yallop et al., 2004). The remaining species of the waterfowl community have been attributed other indicator roles, in some cases as warning signals characterizing specific phases in a process that has been taking place for 40 years (Robledano et al., 2011; Martínez-Fernández et al., 2014).

In the short term (between 2007–2010 and 2018–2019), in a B/A (Before-After) approach that includes the first eutrophic crisis of 2015–2016 (indicators 3 and 4 of Table 2.1), the species that persist in the water mass show changes in number and in their occupation of different sectors of the lagoon (interior and coastal areas). Regarding the summer boat censuses of waterbirds, comparing the postcrisis results with the only reasonably complete preceding series (April–August 2007), remarkable differences are observed in the total abundance of waterbirds and in the numbers of particular species. Total abundance doubled in 2018 compared to the reference period (2007). The species that individually show the greatest increase are two specialist larids *Sterna hirundo* (Common Tern) and *Thalasseus sandvicensis* (Sandwich Tern), strict piscivores able to feed both in shallow waters and in the upper zone of deeper ones. Such species exhibit markedly positive numerical trends as breeders in the area, also during winter in the case of the Sandwich Tern (Ballesteros et al., 2021).

The winter censuses, for their part, show considerable variability within each year, with the common feature that the January census (coincident with the IWC) is not usually the moment of peak abundance. Comparing the average abundance of the years before and after the eutrophic crisis, marked differences occur in the populations of some waterbirds, mainly the strict piscivores, but not in generalists nor in total abundance. Specifically, there seems to be an increase in the Great Crested Grebe (*Podiceps cristatus*) and the Great Cormorant (*Phalacrocorax carbo*), and very notably in the Sandwich Tern (*Thalasseus sandvicensis*), while the Black-necked Grebe (*Podiceps nigricollis*) showed a decrease in numbers.

When applying this B/A approach to the census made from the shoreline (indicator 6), the main responses are somewhat different. These responses were analyzed only for the wintering bird populations, and although the patterns of variation within each year's winter season differ, stands out the change in the mean abundance of two species, *Fulica atra* (Eurasian Coot) and *Mergus serrator* (Red-breasted Merganser), both being less abundant after the first eutrophic crisis. Among the larid species, the largest variation appears in *Thalasseus sandvicensis* (Sandwich Tern), whose abundance is much higher in the postcrisis winter period (2018–2019).

The recent changes have been established from the censuses compiled by Robledano et al. (2019), which correspond to two spatial schemes of ornithological monitoring within the lagoon. They focus respectively on the deep central section of the basin (the already cited indicators 3 and 4) and in the shallow coastal zones (indicator 5). The combination of these spatial approaches may seem useful for evaluating the response of the avifauna in the most recent period (2016 to the present) in which the lagoon condition is clearly differentiated among these two sections: one (the deepest 85%) experiencing the death of all the underwater vegetation and another (the remaining, shallower 15%) where it still survives (Sandonnini et al., 2021). Based on this, Robledano et al. (2019) made comparisons of standardized records (birds/ha) obtained in 2018–2019 (in summer and winter) in the lagoon center and the littoral section, from which it resulted that all species were more abundant in the latter, except the Black-necked Grebe (*Podiceps nigricollis*) whose density was similar throughout all the lagoon surface. More recently, López-Pascual (2023) analyzed the winter census made from the stations of the lagoon's shoreline in 2009–2010 and in 2016–2019, using generalized linear mixed models. Among the responses found (Figure 2.1), we have highlighted the cases of the Black-necked Grebe (*Podiceps nigricollis*) with no significant change in abundance between periods (B/A), but a trend to disperse to areas farther from the shore, and the Great Cormorant (*Phalacrocorax carbo*) that increases significantly in abundance after the eutrophic crisis ($p < 0.001$) and at the same time favors shallower littoral areas. The models comparing periods showed significant decreases in *Fulica atra* and *Mergus serrator* and increases in *Podiceps cristatus, Thalasseus sandvicensis, Larus michahellis* and *Phoenicopterus ruber*. Unfortunately, the sampling coverage in this stage has been fragmentary, and virtually lacking after 2019. In any case, the most recent years with detailed waterbird monitoring have coincided with a period of great ecosystem disorganization (Pérez-Ruzafa et al., 2019), which seems reflected in the bird community. The general trend is a great numerical fluctuation at short time scales, although the time passed since the first eutrophic crisis still seems too short to establish useful conclusions.

Indicators 6 and 7 refer to the land bird communities that characterize the steppe crypto-wetlands peripheral to the Mar Menor. They reflect the decline in populations of the family Alaudidae (main indicators of the original physiognomy of the salt steppe) and the increase in other species favored by the transformation of these habitats towards halophilous scrub and reedbeds. These changes are related to the intensive agricultural transformation of the watershed and the consequent increase of drainage towards these wetlands. Carreño et al. (2008) characterized the effect of these transformations on wetland habitats, essentially an increase in the surface of reedbeds and halophilous scrub at the expense of saline steppe. Robledano et al. (2010) showed the relationship between the surface of saline steppe and the index of abundance of Alaudidae. Using a longer dataset, Guillén et al. (2023) established that the depth of the water table was also a good predictor of the abundance of this family. It is remarkable that the species with the highest conservation value (*Alaudala rufescens*) is precisely one of the ornithological criteria on which the designation of the Mar Menor as EU Bird SPA is based. The fact that this species began to decline in the area shortly before its classification as SPA in 2001 (Robledano et al., 2010, 2019) is an obvious cause for concern.

Indicator number 9 has been chosen because it is the only one related to the maritime border of the lagoon, the *golas* (inlets) that have preserved the historical fishing structures ("encañizadas"). The waterbird censuses carried out in these wetlands, whose microtidal character makes them unique on the Mediterranean coast (Robledano et al., 2018), show great stability and resilience, not being affected in the medium term by the eutrophic crisis of the lagoon. Its own dynamism, governed by maritime currents and storms (Erena et al., 2018), wards off the risk of clogging that has been raised as a justification for its dredging, as a simplistic solution to the phenomenon of eutrophication (ignoring the impact that this could have on the ecosystem and its biodiversity). The high ecological value and sensitivity of this "last semi-natural frontier", together with many other socio-environmental criteria, provided the scientific basis for the evaluation of recreational projects that could threaten its integrity (Robledano et al., 2018). In addition to their intrinsic values, the inlets ("golas") play a crucial role in the restoration of the integrity of the entire lagoon basin. The degree of marine influence and the geomorphology of the inlets regulate colonization rates and genetic exchange with marine waters, determining the spatiotemporal variability and complexity of the ecosystem that represents the basis of its homeostatic capacity (Pérez-Ruzafa et al., 2018). All of this requires great caution in any intervention on these critical elements (García-Oliva et al., 2018).

Finally, indicator 10 is represented in all wetlands on the periphery of the Mar Menor. Among the orders of aquatic birds nesting in the lagoon environment, this small Charadriiform has been selected because it is sensitive to different pressure factors, both those responsible for the structural modification of its habitat (coastal works, urbanization, abandonment of artificial wetlands, landfills) and those related to human use (tourism and recreation). Displaced from its primary habitats of pristine beaches and dunes, and also to a large extent from secondary ones (salt pans and other man-made wetlands), the Kentish Plover has become an indicator of ecological integrity that describes widespread degradation and insufficient management of wetlands. Ballesteros et al. (2021) report a moderate and very significant decline in its breeding population, which, after a regular increase until 2006, has been reduced to almost half its maximum size (Figure 2.1).

Implications for wetland protection, management and monitoring

Wetland birds, whether strict aquatic or terrestrial ones, represent acknowledged bioindicators of ecosystem integrity (Stolen et al., 2004; Ogden et al., 2014; De Zwaan, 2024), depending on their responses to wetland influences and processes (Robledano et al., 2008; Møller & Laursen, 2015; Lehikoinen et al., 2013; Pringle & Burton, 2017). The main changes observed in Mar Menor waterbird bioindicators can be discussed in the framework of the protection and management

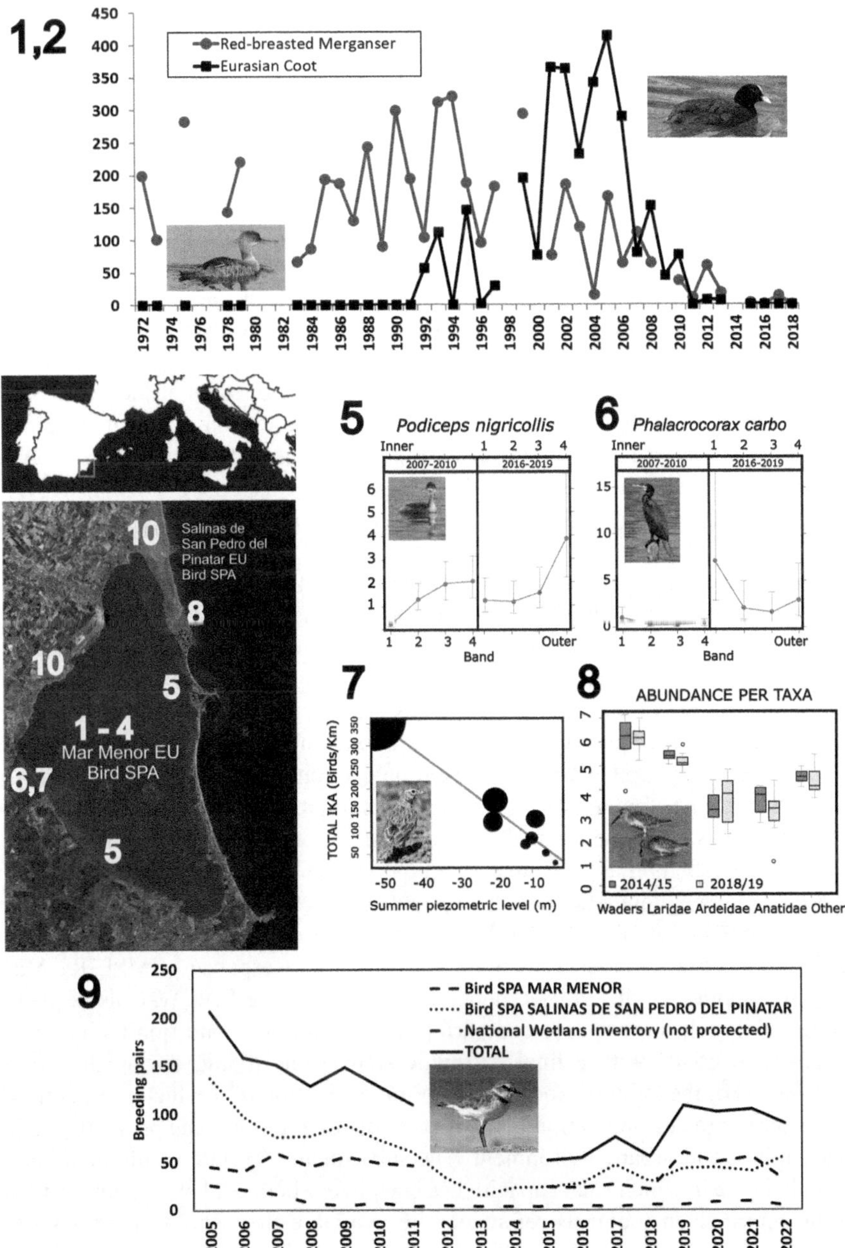

Figure 2.1 Summary of trends in some of the indicators listed in Table 1.1. Indicator numbers are represented approximately in the map of the wetland complex. Indicators 5 and 10 were sampled in locations found along all the lagoon shoreline and in all peripheral wetlands, respectively. Adapted from the works cited in Table 1.1. Bird photos: Wikimedia Commons.

instruments that it has been endowed with, showing the inefficiency of those that ignore the interdependence of wetlands with the dynamics of their terrestrial and maritime areas of influence. Five out of nine bioindicators (1, 2, 6, 7 and 9) have shown negative trends, and most have started close to the date of EU Bird SPA designations.

The Mar Menor Lagoon basin is one of the most intensively monitored aquatic systems, a condition that has been expanded to some areas of its watershed (irrigated areas), where the regional administration is proud of having deployed a network of environmental sensors never attained elsewhere. Biological monitoring of the lagoon is intensive (see e.g. Pérez-Ruzafa et al., 2019, 2023; Fernández-Alías et al., 2022) and numerous research projects have been focused on modeling and interpreting the massive amount of data generated.

Regarding vertebrates, the greatest effort has been directed at the fish assemblages dwelling in the shallow littoral habitats of the lagoon, relevant for different reasons. First, this is the section of the lagoon less affected by the eutrophic crises (phytoplankton blooms), which has allowed macrophyte beds to survive (Zamora-López et al., 2023). Second, they are inhabited by some endangered species like *Aphanius iberus* (Verdiell-Cubedo et al., 2012). And third, this section concentrates most human activities depending for their development on a reasonable environmental quality (bathing, ecotourism), as well as interventions aimed at mitigating environmental degradation (e.g., removal of mud deposits and algal mats, provision of wooden platforms to allow access to water, etc.). Such interventions are not free of controversy, as is expected for any initiative focused at this particularly sensitive part of the lagoon. As a fundamental problem, directly linked to the negative synergies between the eutrophication process and the increasingly frequent extreme climate events, a notable loss of property value has been demonstrated (Lamas Rodríguez et al., 2023) that is associated with a deterioration of the tourist image, both in its conventional modality (sun and beach) and in ecotourism (based on the natural values of the site).

The analysis of bioindicator bird species is necessarily carried out within a Driving Forces, Pressures, State, Impact and Response (FMPEIR; Martínez-Fernández et al., 2007) framework, and under a BA (Wauchope et al., 2022) approach, in that case comparing an oligotrophic condition (prior to the 2015–2016 crisis) with a final eutrophic state. Simplifying, during the previous state (B), the lagoon experienced an intensification of the factors causing or contributing to eutrophication but was still able to assimilate the pressures of the agricultural and urban environment. The subsequent (A) state would be a phase of imbalance in which such capacity has been overwhelmed. Unfortunately, there is no unimpacted (control) reference site available that would allow a BACI approach. The B/A approach enabled by having bird data from before and after the eutrophic crisis is anyway useful but has not had continuity over time. Due to financial constraints, it has not been possible to cover adequately the more recent episodes of eutrophic crisis or mortality of the lagoon fauna, which contrasts with the overinstrumentation and hyper-monitoring of some lagoon parameters by different competent (and sometimes competing) public administrations. It is very convenient that the administration competent in the management of this wetland complex promotes also a much more continuous and intensive monitoring of ornithological indicators.

Concluding remarks

The virtual disappearances of at least two species of wintering birds (Red-breasted Merganser and Eurasian Coot) and the unfavorable trends of others (Short-toed Lark and Kentish Plover) are in themselves signs of a great loss of biological richness. The later respond to changes in land use and in the management of the habitats and landscapes of the lagoon's watershed. The monitoring of the MMSES should take advantage of the indicator value of these and other taxocenoses, most of which have been considered key elements in the management plans of the Natura 2000 sites -Special Protection Areas (SPAs) and Special Areas of Conservation (SACs)-, which unfortunately have taken a long time to be approved (the Comprehensive Management Plan for Natura 2000 areas was published in 2019) and have limited their scope of action to the strict limits of protected areas. The Territorial Land Arrangement Plan for the Mar Menor watershed is still pending approval.

A greater effort is required in monitoring birds, as alert indicators, together with other parameters of the physicochemical state of the lagoon and its watershed. Some studies of the lagoon, which have addressed the changes in water quality parameters in the long term and during recent acute crises (Pérez-Ruzafa et al., 2019), contain interesting references on the role of birds and their habitats in the resilience of the system. First, the restricted connectivity between the lagoon and the Mediterranean Sea, determined by the inlets (*golas*), is recognized as a characteristic that prevents the homogenization of the system and favors a high hydrodynamic and biological heterogeneity, a key functional role that further supports the need to fully preserve these last seminatural frontiers of the lagoon. These studies also show that in the stages prior to the recent crises, the primary production of the Mar Menor has been based mainly on benthic vegetation and its associated biomass of filter feeders, detritivores and scavengers (Pérez-Ruzafa et al., 2019), performing also as an adaptation mechanism to global change (Lloret et al, 2008; Pérez-Martín, 2023). Excess production is accumulated in the sediments or exported from the system through exploitation by migratory fish and waterbirds (Pérez-Ruzafa et al., 2019). From which it is concluded that more studies are also needed on the trophic role of lagoon birds, not only as response indicators, but as components of the system's resilience. The role of sentinels against climate change that has been assigned to coastal lagoons (Brito et al., 2012; Amaral et al., 2023) should be made extensive to aquatic birds, whose alert value in this case extends to all the diversity of natural and seminatural wetlands that are part of the MMSES.

References

Álvarez-Rogel, J., Barberá, G. G., Maxwell, B., Guerrero-Brotons, M., Díaz-García, C., Martínez-Sánchez, J. J., Sallent, A., Martínez-Ródenas, J., González-Alcaraz, M. N., Jiménez-Cárceles, F. J., Tercero, C., & Gómez, R. (2020). The case of Mar Menor eutrophication: State of the art and description of tested nature-based solutions. *Ecological Engineering, 158*, 106086. https://doi.org/10.1016/j.ecoleng.2020.106086.

Amaral, V., Santos-Echeandía, J., Ortega, T., Álvarez-Salgado, X. A., & Forja, J. (2023). Dissolved organic matter distribution in the water column and sediment pore water in a highly anthropized coastal lagoon (Mar Menor, Spain): Characteristics, sources, and benthic fluxes. *Science of The Total Environment, 896*, 165264. https://doi.org/10.1016/j.scitotenv.2023.165264.

Brito, A. C., Newton, A., Tett, P., & Fernandes, T. F. (2012). How will shallow coastal lagoons respond to climate change? A modelling investigation. *Estuarine, Coastal and Shelf Science, 112*, 98–104.

Ballesteros, G. A., Zamora, A., Zamora, J. M., Sallent, A., Hernández, A., Robledano, F., & Fuentes, A. (2021). *Atlas de las Aves Acuáticas del Mar Menor y Humedales de su entorno*. Natursport Ediciones.

Belando, M. D., Ruiz, J. M., García, R., Ramos, A., & García, P. (2014). *Cartografía de la vegetación sumergida del enclave natural de las Encañizadas del Mar Menor*. Instituto Español de Oceanografía and Asociación de Naturalistas del Sureste (ANSE). https://digital.csic.es/bitstream/10261/321084/4/Informe%20Enca%C3%B1izadas%202015.pdf.

Carreño, M. F., Esteve, M. A., Martinez, J., Palazón, J. A., & Pardo, M. T. (2008). Habitat changes in coastal wetlands associated to hydrological changes in the watershed. *Estuarine, Coastal and Shelf Science, 77*(3), 475–483.

Cox, A. R., Frei, B., Gutowsky, S. E., Baldwin, F. B., Bianchini, K., & Roy, C. (2023). Sixty-years of community-science data suggest earlier fall migration and short-stopping of waterfowl in North America. *Ornithological Applications, 125*(4), duad041. https://doi.org/10.1093/ornithapp/duad041.

de Zwaan, D. R., Huang, A., Fox, C. H., Bradley, D. W., & Ethier, D. M. (2024). Occupancy trends of overwintering coastal waterbird communities reveal guild-specific patterns of redistribution and shifting reliance on existing protected areas. *Global Change Biology, 30*(2), e17178. https://doi.org/10.1111/gcb.17178.

Erena, M., Atenza, J. F., Villa, G., García, P., López, J. A., Soria, J. M., Domínguez, J. A., & García, S. (2018). Evolución batimétrica de la gola de la Encañizada en el Mar Menor y su publicación mediante servicios INSPIRE. In *Tecnologías de la información geográfica: Perspectivas multidisciplinares en la sociedad del conocimiento*. Departament de Geografía, Universitat de Valencia. https://congresos.adeituv.es/imgdb//archivo_dpo62588.doc.

Farinós, P., & Robledano, F. (2010). Structure and distribution of the waterbird community in the Mar Menor Coastal Lagoon (SE Spain) and relationships with environmental gradients. *Waterbirds, 33*, 479–493.

Farinós-Celdrán, P., Robledano-Aymerich, F., Carreño, M. F., & Martínez-López, J. (2017). Spatiotemporal assessment of littoral waterbirds for establishing ecological indicators of Mediterranean coastal lagoons. *ISPRS International Journal of Geo-Information, 6*(8), 256. https://doi.org/10.3390/ijgi6080256.

Fernández-Alías, A., Montaño-Barroso, T., Conde-Caño, M. R., Manchado-Pérez, S., López-Galindo, C., Quispe-Becerra, J. I., Marcos, C., & Pérez-Ruzafa, A. (2022). Nutrient overload promotes the transition from top-down to bottom-up control and triggers dystrophic crises in a Mediterranean coastal lagoon. *Science of the Total Environment, 846*, 157388. https://doi.org/10.1016/j.scitotenv.2022.157388.

García-Oliva, M., Pérez-Ruzafa, Á., Umgiesser, G., McKiver, W., Ghezzo, M., De Pascalis, F., & Marcos, C. (2018). Assessing the hydrodynamic response of the Mar Menor lagoon to dredging inlets interventions through numerical modelling. *Water*, *10*(7), 959. https://doi.org/10.3390/w10070959.

Garcia-Pintado, J., Martínez-Mena, M., Barberá, G. G., Albaladejo, J., & Castillo, V. M. (2007). Anthropogenic nutrient sources and loads from a Mediterranean catchment into a coastal lagoon: Mar Menor, Spain. *Science of the Total Environment*, *373*(1), 220–239.

Guaita-García, N., Martínez-Fernández, J., Barrera-Causil, C. J., & Esteve, M. A. (2021). Local perceptions regarding a social-ecological system of the Mediterranean coast: The Mar Menor (Región de Murcia, Spain). *Environment, Development and Sustainability*, *23*, 2882–2909. https://doi.org/10.1007/s10668-020-00697-y.

Guillén, E., Carreño, M. F., & Robledano, F. (2023). Cambios hidrológicos y biodiversidad: el declive de la avifauna esteparia del entorno del Mar Menor. In Sánchez Gallardo, L. (Coord.), *Mirando a los ríos desde el mar. Viejos y nuevos debates para una transición hídrica justa* (pp. 90–95). Editum. Ediciones de la Universidad de Murcia; Fundación Nueva Cultura del Agua. https://doi.org/10.6018/editum.3003.

Hernández-Gil, V., & Robledano-Aymerich, F. (1997). La comunidad de aves acuáticas del Mar Menor (Murcia, SE de España): aproximación a su respuesta a las modificaciones ambientales en la laguna. In *Actas de las XII Jornadas Ornitológicas Españolas: Almerimar (El Ejido-Almería), 15 a 19 de septiembre, 1994* (pp. 109–121). Instituto de Estudios Almerienses. https://dialnet.unirioja.es/servlet/fichero_articulo%3Fcodigo=2244582%26orden=78933.

Krause-Jensen, D., Sagert, S., Schubert, H., & Boström, C. (2008). Empirical relationships linking distribution and abundance of marine vegetation to eutrophication. *Ecological Indicators*, *8*(5), 515–529.

Lamas Rodríguez, M., Garcia Lorenzo, M. L., Medina Magro, M., & Perez Quiros, G. (2023). Impact of climate risk materialization and ecological deterioration on house prices in Mar Menor, Spain. *Scientific Reports*, *13*(1), 11772. https://www.nature.com/articles/s41598-023-39022-8.

Lehikoinen, A., Jaatinen, K., Vähätalo, A. V., Clausen, P., Crowe, O., Deceuninck, B., Hearn, R., Holt, C. A., Hornman, M., Keller, V., Nilsson, L., Langendoen, T., Tománková, I., Wahl, J., & Fox, A. D. (2013). Rapid climate driven shifts in wintering distributions of three common waterbird species. *Global Change Biology*, *19*(7), 2071–2081.

Lloret, J.; Marín, A., & Marín-Guirao, L. (2008) Is coastal lagoon eutrophication likely to be aggravated by global climate change? *Estuar. Coast. Shelf Sci*, *78*, 403–412

López-Pascual, J. J. (2023). *La avifauna acuática del Mar Menor antes y después de la crisis eutrófica*. Final Degree's Thesis. University of Murcia.

Martínez-Fernández, J., Esteve, M. A., Martínez-Paz, J. M., Carreño, F., Robledano, F., Ruiz, M., & Alonso, F. (2007). Simulating management options and scenarios to control nutrient load to Mar Menor, *Southeast Spain. Transitional Waters Monographs. TWM, Transitional Waters Monographs*, *1*, 53–70.

Martínez-Fernández, J., Esteve-Selma, M. A, Martínez-Paz, J. M., Carreño-Fructuoso, M. F., Martínez-López, J., Robledano, F., & Farinós, P. (2014). Trade-offs between biodiversity conservation and nutrients removal in wetlands of arid intensive agricultural basins: The Mar Menor Case, Spain. *Developments in Environmental Modelling*, *26*. https://doi.org/10.1016/B978-0-444-63249-4.00012-9.

Martínez Fernández, J., Esteve Selma, M. A. (2020). El colapso ecológico de la laguna del Mar Menor. In: *Retos de la planificación y gestión del agua en España. Informe 2019*. Observatorio de las Políticas del Agua (OPPA). Fundación Nueva Cultura del Agua. pp. 61–75.

Møller, A. P., & Laursen, K. (2015). Reversible effects of fertilizer use on population trends of waterbirds in Europe. *Biological Conservation*, 184, 389-395.

Moss, B. (2015). Biodiversity climate change impacts report card technical paper. Freshwaters, climate change and UK conservation. *UK Research and Innovation.* https://www.ukri.org/wp-content/uploads/2021/12/101221-NERC-LWEC-InfrastructureReportSource17-FreshwatersConservation.pdf.

Noordhuis, R., van der Molen, D. T., & van den Berg, M. S. (2002). Response of herbivorous water-birds to the return of Chara in Lake Veluwemeer, The Netherlands. *Aquatic Botany*, *72*(3–4), 349–367.

Ogden, J. C., Baldwin, J. D., Bass, O. L., Browder, J. A., Cook, M. I., Frederick, P. C., Frezza, P. E., Galvez, R. A., Hogdson, A. B., Meyer, K. D., Oberhofer, L. D., Paul, A. F., Fletcher, P. J., Davis, S. M., & Lorenz, J. J. (2014). Waterbirds as indicators of ecosystem health in the coastal marine habitats of southern Florida: 1. Selection and justification for a suite of indicator species. *Ecological indicators*, *44*, 148–163.

Pavón-Jordán, D. (2017). *Waterbirds in a changing world: Effects of climate, habitat and conservation policy on european waterbirds.* Doctoral Dissertation, Faculty of Biological and Environmental Sciences of the University of Helsinki. https://helda.helsinki.fi/handle/10138/180162.

Pérez-Martín, M. Á. (2023). Understanding nutrient loads from catchment and eutrophication in a salt lagoon: The Mar Menor case. *Water*, *15*(20), 3569. https://doi.org/10.3390/w15203569.

Pérez-Ruzafa, A., Marcos, C., & Pérez-Ruzafa, I. M. (2018). When maintaining ecological integrity and complexity is the best restoring tool: the case of the Mar Menor lagoon. *Management and Restoration of Mediterranean coastal lagoons in Europe*, *10*, 67–95. https://tinyurl.com/2ylohprl.

Pérez-Ruzafa, A., Campillo, S., Fernández-Palacios, J. M., Garcia-Lacunza, A., García-Oliva, M., Ibañez, H., Navarro-Martínez, P. C., Pérez-Marcos, M., Pérez-Ruzafa, I. M., Quispe-Becerra, J. I., Sala-Mirete, A., Sánchez, O., & Marcos, C. (2019). Long-term dynamic in nutrients, chlorophyll a, and water quality parameters in a coastal lagoon during a process of eutrophication for decades, a sudden break and a relatively rapid recovery. *Frontiers in Marine Science*, *6*, 26. https://doi.org/10.3389/fmars.2019.00026.

Pérez-Ruzafa, A., Dezileau, L., Martínez-Sánchez, M. J., Pérez-Sirvent, C., Pérez-Marcos, M., von Grafenstein, U., & Marcos, C. (2023). Long-term sediment records reveal over three thousand years of heavy metal inputs in the Mar Menor coastal lagoon (SE Spain). *Science of the Total Environment*, 166417. https://doi.org/10.1016/j.scitotenv.2023.166417.

Perrow, M. R., Schutten, J. H., Howes, J. R., Holzer, T., Madgwick, F. J., & Jowitt, A. J. D. (1997). Interactions between coot (*Fulica atra*) and submerged macrophytes: the role of birds in the restoration process. *Hydrobiology*, 342, 141–255. https://doi.org/10.1007/978-94-011-5648-6_26.

Pringle, H. E. K., & Burton, N. H. K. (2017). Improving understanding of the possible relationship between improving freshwater and coastal water quality and bird interest on designated sites–phase 1 review. *BTO* Research Report, (696). https://www.bto.org/sites/default/files/publications/rr696.pdf.

Robledano, F., Pagán, I., & Calvo, J. F. (2008). Waterbirds and nutrient enrichment in Mar MenorLagoon, a shallow coastal lake in southeast Spain. *Lakes and Reservoirs: Research and Management*, *13*, 37–49.

Robledano, F., Esteve, M. A., Farinós, P., Carreño, M. F., & Martínez-Fernández, J. (2010). Terrestrial birds as indicators of agricultural-induced changes and associated loss in conservation value of Mediterranean wetlands. *Ecological Indicators*, *10*(2), 274–286. https://doi.org/10.1016/j.ecolind.2009.05.006.

Robledano, F., Esteve, M. A., Martínez, J., & Farinós, P. (2011). Determinants of wintering waterbird changes in a Mediterranean coastal lagoon affected by eutrophication. *Ecological Indicators*, *11*, 395–406. https://doi.org/10.1016/j.ecolind.2010.06.010.

Robledano, F., Esteve, M. A., Calvo, J. F., Farinós, P., & Carreño, M. F. (2018). Multicriteria assessment of a proposed ecotourism, environmental education and research infrastructure in a unique lagoon ecosystem: The Encañizadas del Mar Menor (Murcia, SE Spain). *Journal for Nature Conservation*, *43*, 201–210.

Robledano, F., Zamora, A., Farinós, P., León, M., Martínez, J., García, Castellanos F. A., & García Meseguer, A. J. (2019). *Monitorización de la avifauna y otras especies indicadoras del estatus medioambiental y el estado de alteración de la biodiversidad como consecuencia del proceso de eutrofización del Mar Menor.* Report for the Mar Menor Scientific Advisory Committee by contract between TRAGSATEC-University of Murcia.

Ruiz-Fernández, J. M., Belando-Torrentes, M. D., Bernardeau-Esteller, J., & Mercado-Carmona, J. M. (2022). Mar Menor lagoon: an iconic case of ecosystem collapse. *Harmful Algae News*, *70*, 1–5. https://www.e-pages.dk/ku/1544/.

Sandonnini, J., Ruso, Y. D. P., Melendreras, E. C., Barberá, C., Hendriks, I. E., Kersting, D. K., & Casalduero, F. G. (2021). The emergent fouling population after severe eutrophication in the Mar Menor coastal lagoon. *Regional Studies in Marine Science*, *44*, 101720. https://doi.org/10.1016/j.rsma.2021.101720.

Stolen, E. D., Breininger, D. R., & Frederick, P. C. (2004). Using waterbirds as indicators in estuarine systems: Successes and perils. In Bortone, S. A. (Ed.), *Estuarine Indicators* (pp. 209–412). CRC Press.

Verdiell-Cubedo, D., Torralva, M., Andreu-Soler, A., & Oliva-Paterna, F. J. (2012). Effects of shoreline urban modification on habitat structure and fish community in littoral areas of a Mediterranean coastal lagoon (Mar Menor, Spain). *Wetlands*, *32*, 631–641.

Vidal-Abarca, M. R., Esteve Selma, M. A., & Suárez Alonso, M. L. (2003). *Los Humedales de la Región de Murcia: Humedales y Ramblas de la Región de Murcia*. Murcia: Consejería de Agricultura, Agua y Medio Ambiente.

Wauchope, H. S., Jones, J. P. G., Geldmann, J., Simmons, B. I., Amano, T., Blanco, D., Fuller, R. A., Johnston, A., Langendoen, T., Mundkur, T., Nagy, S., & Sutherland, W. J. (2022). Protected areas have a mixed impact on waterbirds, but management helps. *Nature*, 605 (7908), 103–107.

Yallop, M. L., O'Connell, M. J., & Bullock, R. (2004). Waterbird herbivory on a newly created wetland complex: potential implications for site management and habitat creation. *Wetlands Ecology and Management*, *12*, 395–408.

Zamora, A. (2015). *Aves acuáticas como indicadores para la gestión de sistemas mareales mediterráneos: el caso de las Encañizadas del Mar Menor (Murcia, SE España)*. Trabajo Fin de Máster (Máster en Gestión de la Biodiversidad en Ambientes Mediterráneos), Universidad de Murcia.

Zamora-López, A., Guerrero-Gómez, A., Torralva, M., Zamora-Marín, J. M., Guillén-Beltrán, A., & Oliva-Paterna, F. J. (2023). Shallow waters as critical habitats for fish assemblages under eutrophication-mediated events in a coastal lagoon. *Estuarine, Coastal and Shelf Science*, *291*, 108447. https://doi.org/10.1016/j.ecss.2023.108447

3 Agrarian reason, deep histories and environmental destruction in the Campo de Cartagena-Mar Menor[1]

A Gramscian world-ecology perspective

Andrés Pedreño Cánovas, Carlos de Castro Pericacho and Miguel Ángel Sánchez García

Introduction

A critique of the political economy perspective of the agrarian-intensive model of the Region of Murcia frames this chapter and, at the same time, is intertwined with *a Gramscian spirit*.

On the one hand, from the ecology-world approach developed by Jason W. Moore (2020), we consider that the process of capitalisation developed by the intensive horticultural model in the Region of Murcia, which is in the area of Campo de Cartagena one of its central geographies, implies a certain organisation of a historical nature to enable the appropriation of cheap resources (water, soil, human labour, etc.). The environmental problems of the Mar Menor Lagoon emerge from this historical nature.

On the other hand, we adopt a Gramscian perspective to understand that the agrarian problem of the Region of Murcia, as it is nowadays formulated, requires a historical key conceptualised by the Sardinian theorist and politician in terms of "the southern question". In the Spanish reality, too, a southern question was formed, of which the problem of the historical underdevelopment of the regions of southeastern Spain (Alicante, Murcia and Almería) took part.

The southern question, as Gramsci taught us to think of it, "does not consist only in the normal historical relations between town and country, created by the development of capitalism in all the countries of the world" (Gramsci, 2024/1930, p. 233). It is not the "peasant and agrarian question in general" but it is historically determined by the way in which the development of capitalism took place in Italy. Thus, the Sardinian thinker thought of the southern question as a territorial question: an expression of Italy's unequal development between the industrial north and the peasant south, that is, as a form of exploitation of the southern rural masses by the industrial-modernising bloc through a protectionist policy that harmed southern Italy and only benefited the industrial north. Gramsci put all his political efforts into defending, within the Communist Party of Italy, the political work of building

DOI: 10.4324/9781003489078-4

an alliance between the industrial proletariat of the North and the poor peasantry of the South as a strategy for overcoming unequal development and exploitation which, at the same time, would make the fascist solution to the territorial problem unviable.

In Spain, we also find a series of thinkers who raised the agrarian question in terms of national territorial articulation, and even as part of the political regeneration programme. The classic texts of Joaquín Costa, such as "Oligarchy and caciquism as the current form of government in Spain" of 1901, among others, are emblematic of this type of approach (Costa, 1984). The problem of water distribution among the different Spanish southern regions to make irrigated agriculture possible was considered, from different political positions (regenerationism, federalism, republicanism, socialism, national-catholicism, Francoism), as a social issue to solve the situation of the poor peasantry and also as a material basis for the territorial articulation of the nation.

The environmental tensions that float in the greenish waters of the Mar Menor Lagoon reflect the historical threads that bind together the policies that finally addressed the issue of the southeast, the deployment of the Murcian model of intensive agriculture and the deep history of the farmers who transformed "the desert" (that is, the semi-arid territories and rain-fed fields of the southeastern regions) into an orchard. The waters of the Mar Menor tell a choral story. The purpose of this chapter is to analyse this bundle of connections.

The creation of an area of capitalisation for fresh fruit and vegetable production and the new class differentiation in the countryside

From the beginning of the 20th century, a cycle of fruit and vegetable production began in the fertile plains of the Segura, specialising early on in an eminently export-oriented vocation. This expansion of commercial agriculture went hand in hand with an important development of the vegetable canning industry, which became an industrial production system specifically linked to the Segura Valley. This economic activity boosted the demography and the urbanisation process. By means of the techniques for raising water from the fertile plains and with the exploitation of the aquifers (Pérez Picazo and Lemeunier, 1990), the unirrigated areas immediately connected to the traditional market gardens were transformed into irrigated land. This dynamism of the fruit and vegetable sector led to a significant increase in the percentage of land dedicated to fruit and vegetable crops from 19.7% in 1900 to 38.85% in 1935 (Pérez Picazo and Lemeunier, 1990, p. 232).

A geography of modernisation, still in its infancy, began to take shape, whose demands for new water resources would be met through the exploitation of groundwater. But these water resources proved to be insufficient for this new cycle of accumulation, which led the "modernising social bloc", made up of landowning oligarchies, local political elites and the agricultural and industrial bourgeoisie, to demand that the State build reservoirs to regulate the Segura basin. This marked the beginning of the phase of the "Great Hydraulics" which would extend throughout the 20th century with the deployment of a succession of reservoirs and dams which would achieve the full artificialisation of the Segura River basin. Paradoxically, the hydrographic policy initiated during the years of the Republic ended up becoming one of the symbols of the economic modernisation of the Franco dictatorship, synthesised in the frenetic construction of reservoirs throughout the length and breadth of the country. In fact, in 1953, the publication of a decree for the regulation of irrigation of the Segura River facilitated the construction of the network of reservoirs and the legalisation of previously illegal small irrigation schemes (Pérez Picazo and Lemeunier, 1990, p. 233; Martínez Carrión, 2002, p. 488). The culmination of the great hydraulic project was the construction of the Tajo-Segura water transfer at the end of the 1970s.

The southern question in southeastern Spain, which was expressed as in other regions as the question of the large property and the poor peasantry, was reinterpreted as a question of lack of water by the modernising-oligarchic bloc, that is, as a need for water infrastructures to underpin the trend towards the formation of intensive and industrial agriculture. The Franco dictatorship, with the development of hydraulic works and the Development Plan of Minister López Rodó in 1962, made possible an oligarchic response to the underdevelopment of the peripheral regions, albeit in a very unequal manner with respect to the southeast of Spain. The southeast was practically excluded from the policy of encouraging development poles, disconnected from the main communication routes and emigration abroad continued to be the people's response to underdevelopment: "only the Tajo-Segura water transfer, which would be resumed two years later, deploying the plans of Indalecio Prieto, would slightly improve the prospects of the countryside of Murcia and Cartagena, which would eventually reach a great export potential" (Villacañas, 2022, p. 330).

Gramsci asked himself in 1920 on the Italian southern question, "in whose favour should the power of the state which has unified and centralised the process of production be resolved, in favour of the plutocracy or in favour of the working class?" and asserted that "it is necessary for the working class to take the power of the state into its own hands in order to resolve it in its favour and in favour of the poor peasants" (Gramsci, 2024/1930, p. 165). In order to subdue "capitalism to useful labour", as Gramsci advocated, the working class will take "industry and banking" into its hands and in this way

> the proletariat will resolve the enormous power of state organisation to support the peasants in their struggle against the landlords, against nature and against misery. In this way it will give credit to the peasants, it will institute the cooperative, it will guarantee security of person and property against plunderers, it will carry out the public work of sanitation and irrigation. It will do all this because its interest is to increase agricultural production, to have and preserve the solidarity of the peasant masses, to convert industrial production into useful work of peace and fraternity between town and country, between North and South.
>
> (Gramsci, 2024/1930, pp. 160–161)

However, the "passive revolutions"[2] of fascism in Italy and Francoism in Spain resolved the southern question in favour of "the plutocracy" and, therefore, in a way far removed from Gramsci's communist project of a worker-peasant alliance. Franco's dictatorship laid the foundations for the formation of a space of capitalisation in southeastern Spain through the development of intensive agriculture. From the 1970s onwards, the great transformation to irrigation was driven by the extraction of groundwater and, above all, by the arrival of water from the Tajo-Segura water transfer and the land consolidation plans of the National Institute for Agrarian Reform and Development. Spain's entry into the European Economic Community in 1986 and the Single European Market in 1991 opened a large market for fruit and vegetables from the southern regions. This laid the foundations for the current "differentiated fruit and vegetable model" (Segura and Pedreño, 2006) characterised by:

1 A high degree of productive specialisation in fruit and vegetables: more than 70% of the Final Agricultural Production of the Region of Murcia is fruit and vegetables. Specifically, Campo de Cartagena specialises mainly in leafy vegetables, although it also specialises in citrus fruits.
2 The incorporation of technological and organisational innovations to promote productive intensification that allows for a constant increase in physical and economic productivity based on varietal specialisation and reconversion, as well as productive intensification through the technological package of fertigation. Counter-seasonal production enables the successful insertion of Murcia's agricultural territory into global agricultural chains, over which the large food distribution groups exercise a high degree of control. In this way, we can affirm that a model of agro-industrial development that is "endogenous, but externally oriented" by the large global supermarkets is being consolidated.
3 The expansion or extension of the occupied areas. From the progressive transformation of the immense stock of unirrigated land existing in the region, as a privileged way of increasing productive capacity through the expansion of cultivated areas, a process of centralisation and concentration of capital has taken place which has made it possible for the Murcian agricultural model to be based on large-scale production units. The constant growth in demand has led to a continuous and renewed demand for land and nature for its capitalisation.
4 The centralisation of production, through the constitution of farms with a high territorial and technical-economic dimension and the formation of business structures integrating farms and activities under a single management unit. This explains the significant growth and expansion of large capital companies, both regional and foreign (national and foreign), which take the lead in agro-industrial development. This led to a socio-economic dynamic of progressive displacement and expulsion of small farms, which was only counteracted by the cooperative societies, which increased their number and that of the small- and medium-sized farms grouped together in them, providing them with a way out that allowed for relative stabilisation and the overcoming of the crisis to which they seemed to be condemned. The cooperatives promoted by small- and

medium-sized farmers finally developed, not under the conditions that Gramsci advocated, but under conditions of global capitalism, although the support of the public policy of the new autonomous communities created by the democratic constitution of 1976 was decisive.

5 Productive integration according to a "forward growth" model, from agricultural activity to processing and distribution. This implies overcoming the subordination of agriculture to other activities, the capture of added value and the centrality of the agricultural processing warehouse – for its conversion into a product-market – as it assumes the strategic function of directing and coordinating the whole process of cultivation-production-processing and marketing. This implies the constitution not only of production units of a high territorial dimension but also, especially, of a technical-economic dimension, giving rise to the formation of complex business structures that integrate various farms and activities under a single management unit. At the same time, the need of European supermarket chains to minimise the time between harvesting, packaging and delivery to obtain greater freshness and quality of the products has favoured the development of a specialised cold logistics in the Region. Organised on the basis of quality standards and *just-in-time* principles, this export logistics is concentrated in large regional transport and storage companies that in recent years have been incorporating new logistics services to the classic long-distance transport offer.

6 The extension of wage labour. This is a production model that is highly intensive in salaried labour, both in fieldwork (planting, harvesting and other specialised tasks) and in the handling process in the warehouse. At first, this need was covered by the traditional pools of day labourers historically rooted in the southern regions, and later, as Spanish economic modernisation transferred this population of rural workers to other sectors of the economy (construction, hotels and restaurants, etc.), it was international migrations that made it possible to build up a large reserve pool of labour to cover both the growing demand for labour and to ensure a surplus available to cover peak periods of activity.

In short, the Murcian production model entails the definitive cancellation of the old agrarian class structure of the Gramscian southern question (large rentier landowners and poor peasantry). With industrial and intensive agriculture, a new and more differentiated agrarian class structure emerged, made up of, firstly, the capital class of large, wealthy agrarian entrepreneurs with the capacity to accumulate productive resources, generate economies of scale and participate in expanded capitalist reproduction; secondly, other capital classes linked to extra-agrarian sectors, such as industry or finance, among which commercial capital (the large food distributors that articulate global fruit and vegetable chains) should be highlighted; thirdly, the small- and medium-sized farmers who have managed to counteract the expulsion tendencies by setting up cooperatives and, to the extent that they have been successful in their capacity for accumulation and reproduction, have managed to stabilise themselves as a differentiated class and contradictory position between the capital and labour classes; and finally the new agricultural wage earners, also differentiated in terms of gender and ethno-national origin, and who specifically make up the labour class.

Social condition of possibility of capitalisation: appropriation of cheap nature

The Region of Murcia has emerged as an area specialising in the production of quality foodstuffs on the basis of an adequate endowment of comparative advantages in terms of natural resources, especially soil and climate, and labour at a limited cost, which has led to its specialisation in production orientations with high biological and economic yields, obtained by means of meticulously rationalised capital and labour intensive production processes, thus enabling the consolidation of the exporting vocation of the fruit and vegetable model.

This model of intensive agriculture implies a drastic disengagement with nature and territory. This process of deterritorialisation, in turn, entails a new social production of nature and a new social production of space, that is, a relative space[3] elaborated by a series of social and productive practices specific to the intensive horticultural model, which will enable its insertion into global commodity chains, especially with the constitution of the Single European Market at the beginning of the 1990s.

The frontiers of appropriation of nature have thus been defined as conditions of possibility for agro-industrial development in the Region of Murcia. In other words, the availability of cheap nature (mainly extra-human labour, but also human) is a condition for the capitalisation experienced by the agricultural areas of the Region of Murcia. These frontiers of appropriation can be summarised as follows:

1 Appropriation of drylands, woodlands and natural areas for their transformation into cultivation areas

   This cheap appropriation frontier has been decisive in ensuring that the availability and price of land are not limiting factors, making possible continuous increases in the productivity of capital. The existence of an important pool of transformable land at low prices – due to factors such as abundance, the inexistence of more profitable alternatives and the possibility of resorting to leasing as an alternative – has caused land – as a factor of production – to undergo a metamorphosis, from its original and intrinsic character as an appropriable natural resource, determined by the degree of intrinsic fertility, to that of "produced capital" by means of physical interventions – artificial creation of useful soil by means of clearing, ploughing, fanning, nutrient supply, etc. – and economic, since these actions are the result of the cumulative application of capital and labour. This phenomenon of artificial soil creation has been carried out by large corporate farms specialising in intensive crops and has developed generically throughout the Region, although it is predominant in areas such as Campo de Cartagena-Mar Menor, Vale de Guadalentín and Águilas-Mazarrón.

2 Appropriation of groundwater

   On August 2, 1968, Franco's dictatorship approved the construction of the Tajo-Segura aqueduct (Martínez Carrión, 2002, p. 488). The start of work on the aqueduct from the Tajo River had the effect of generating expectations among farmers which led to a significant growth in irrigation between 1974 and 1981, reaching 187,974 ha in 1990, almost double the figure for 1970

(Martínez Carrión, 2002, p. 491). This expansion was fundamentally based on the exploitation of aquifers, given the insufficiency of the basin's own surface resources. When, at the beginning of the 1980s, and coinciding with a period of drought, it became clear that the expectations placed in the Tajo-Segura aqueduct were not being fulfilled after it came into operation, the water deficit became structural and remained a constant until the present day.[4] The intensive exploitation of the aquifer was facilitated by the application of new well drilling techniques for the extraction of water from the subsoil. From 1950 to the mid-1970s, the contribution of the wells was at levels close to 200 hm^3 on average per year, while from that moment onwards an irregular increase began, reaching 500 and 600 hm^3 in some years (Martínez Carrión, 2002, p. 489).

Both the water in the aquifers and the water from the transfer are considered a public asset and the Confederación Hidrográfica de la Cuenca del Río Segura is the state authority responsible for its management. Individual or collective owners (Comunidades de Regantes) are entitled not to ownership but to the use of the water by means of an administrative concession from the *Confederación*. However, the Confederation lacks the means to control compliance, which is why clandestine wells are frequently opened and authorised aquifers overexploited (Fanlo, 2002). For example, in the case of the Segura basin, in 2001, there were an estimated 20,350 wells for irrigation in the Segura basin, a figure that contrasts with the 4,500 wells authorised by the *Confederación* (Baños et al., 2009, p. 89).

The water deficit in the region is therefore structural and is mainly due to agricultural use. The *Confederación* calculates that for the period 1980–2006, the average annual contribution of natural surface water resources, mainly from reservoirs, was 704 hm^3, while the average annual contribution of natural groundwater resources was 546 hm^3. This water availability of an annual average of 1,250 hm^3 contrasts with the estimated demand in 2010 of 1,779 hm^3, of which 86.6% corresponded to agricultural uses, 10.7% to supplying the population and the rest to industrial uses, services, and maintenance of wetlands. In short, the collective strategy to try to solve the water deficit, together with the permanent demand of the irrigation-modernising block for new water transfers from the Tajo River, has consisted of the intensive extraction of groundwater. It could be said that intensive irrigation in the Campo de Cartagena, as in other areas such as the Guadalentín or Ascoy-Sopalmo, is based more on the extraction of groundwater than on water from the Tajo-Segura water transfer. The environmental collapse of the Mar Menor is closely related to this dynamic of intensive groundwater extraction from the Campo de Cartagena aquifer by intensive agriculture.

3 Appropriation of nature for its conversion into a sink for agro-waste: the Mar Menor Lagoon.

Some scientific studies on the Mar Menor already diagnosed an incipient process of eutrophication in the 1980s, after the first transformations to irrigation in the Campo de Cartagena resulting from the arrival of the Tajo-Segura water transfer. From the mid-90s onwards, the invasions of jellyfish started to be recurrent during the summer season. This fact, together with other symptoms, raised the alarm among the scientific community about the possibility that a process of eutrophication was underway, caused by the entry of nutrients from agriculture in the Campo de Cartagena. Eutrophication is a biophysical process that is difficult to grasp and requires knowledge, such as ecology, specialised in establishing relationships between environmental "facts". It was in 2000, with the application by Martínez and Esteve (2000) of a first dynamic model of irrigation in the Campo de Cartagena and the export of fertilisers, that the effects on the natural environment and other socio-economic uses began to be glimpsed. Already then, this methodology allowed us to corroborate that

> the substantial increase in the contribution of nutrients to the Mar Menor through the ravine, subsurface drainage and very recently the dumping of brine, has begun to alter the oligotrophic character that its waters have always had through eutrophication, although very initial but progressive.
> (Martínez and Esteve, 2000)

and even related the "summer jellyfish crises with the nutrient inputs from agriculture in the Campo de Cartagena". What in 2000 was "a very initial but progressive eutrophication", in 2016, the so-called "green soup" revealed that eutrophication was already a very advanced phenomenon caused by a phytoplankton *bloom.*

While the evidence linking the eutrophication process with nutrients from chemical fertilisation of agriculture in Campo de Cartagena was taking shape, European legislation on the regulation of agricultural pollution by nitrates was being transposed into Spanish law (Directive 91/676/EEC, which was transposed into Spanish law by R.D. 261/1996). Under this legislation, the Campo de Cartagena was declared a Vulnerable Zone in 2001 for generating agricultural nitrate pollution, and in 2003, the code of good agricultural practices and action programmes to reduce nitrate pollution were approved (in 2003, 2009, 2011 and 2016). In view of the phytoplankton bloom of 2016 and 2017, public regulation of nitrate pollution has indeed been totally ineffective. This was further aggravated by the discharge of brine, with high nitrate concentrations, from the hundreds of illegal desalination plants that farms had installed in recent years to provide water for irrigation. The exponential growth of illegal irrigation, which, according to a recent report by ANSE (2018), is around 12,000 ha, 25% of all irrigation in the Campo de Cartagena, would not have been possible without the approval of the *Confederación Hidrográfica del Segura*, which did not exercise its powers to control the water used for irrigation on farms, the drilling of wells and the discharge of brine into the aquifer. The sink for the waste (brine and nitrates) has been the Mar Menor Lagoon.

4 Appropriation of cheap labour supplied by international migrations

In intensive agriculture, capital and migratory flows are closely intertwined. International migrations have covered the growing need for salaried labour required to grow crops (planting, harvesting and other tasks) and the handling or packaging of produce in warehouses. Today, the immigrant status of most agricultural wage earners is a fact (Pedreño and Riquelme, 2022). The number of workers who have arrived through migratory flows and their diversity of origins have given these territories a characteristic cosmopolitanism. The vulnerability of this labour force has also been an important mechanism of wage devaluation in fruit and vegetable production enclaves (Pedreño, 1999). The figure of the "poor worker" has been a permanent feature of the way in which immigrant workers have been inserted into the labour market in the Murcian countryside. This has posed obvious problems of social integration.

The access to and use of migrant labour allowed the intensification of crop cultivation, with all that this implies in terms of water use, fertilisers and the dumping of waste into the Mar Menor. In other words, the exploitation of migrant labour was the necessary condition for the intensification of water and land appropriation. The scale of the appropriation of nature could have been smaller without the domination over migrant workers.

The deep story: from desert to irrigated farmer

Along with the analysis of the objective characteristics of the fruit and vegetable model in Murcia, we also need an approach to the subjectivity of the farmers to understand their collective stance on the environmental issue raised by the degradation of the Mar Menor. To this end, the concept of "deep story", defined by Arlie Hochschild (2016) to investigate the worldview of the American right, is particularly productive to capture that story about "what one feels, the story told by feelings using a language of symbols and eliminating the rational".

Everything suggests that the fears and expectations of the southern question, that is, a problematic relationship with aridity in terms of fears of drought, the absence of water, floods, etc., are palpable in this deep history. Thanks to the presence of perennial watercourses, the Segura River and its tributaries, the inhabitants of these lands cultivated the fertile plains and gave rise to a characteristic system of market gardens, through what water historians call the long period of "traditional hydraulics", which began around 1450 and lasted for almost four centuries, that is, until the mid or late 19th century (Pérez Picazo and Lemeunier, 1994). During this period, the agricultural landscape of the Segura basin, and specifically that of the Region of Murcia, was structured in the form of a "geography of contrast", in which a socio-spatial differentiation is established between *la huerta* (the orchard) (on the axis of the Segura River and its tributaries) and *el campo* (the countryside) (the vast dry land area which extends outside the perennial watercourses of the Segura basin).

The *huerta* concentrated the demographic and economic dynamism of the region. The possibility of capturing water from the river by means of diversion dams or weirs and irrigation canals made it possible to develop a series of commercial crops and, especially, the silk industry. But this logic of demographic and economic concentration along the Segura River meant, at the same time, exposing these populations to the risk of flooding, which would occur periodically with certainly catastrophic effects, without freeing them from the fear of drought. The control of water for irrigation and for supplying urban centres did not prevent it from becoming uncontrollable in nature.

In the *campo*, the problem of aridity and drought was particularly acute. This is why in this area, the populations and economic activity were located around the water points (with a triple origin: rain, underground and spring), according to a logic that could be rigorously defined as an oasis. The importance of scattered settlements and the development of crops that are not very demanding in terms of water, such as cereal growing, reflect social strategies that adapt to situations of scarcity imposed by aridity. The yields of these rain-fed crops were hardly enough to guarantee the subsistence of peasant families, which generated a population surplus that was forced to emigrate.

This "geography of contrast" was, therefore, shaped by a generous nature – in the fertile plains irrigated by the Segura and its tributaries – or by a sober one – in the dry fields – but in either case, this nature was perceived as uncertain and unpredictable. It was these collective representations that were to be permanently mobilised during the long and tortuous modernisation of the lands of southeastern Spain throughout the 20th century. Aridity, and its inherent risks of flooding and drought, limited social and economic development.

These fears, secular in the southeast, were to define the development intervention policy of Franco's dictatorship. This logic of spatial and water rationalisation of the Segura basin was assumed as an authentic work of the State, which acted in the basin through development plans, land consolidation plans and the works of the *Confederación*, according to the subjective needs expressed by the agrarian-modernising bloc and which were synthesised in the slogan that presided over the meetings of the Southeast Interprovincial Trade Union Economic Council: "The Spanish Southeast will never have development as long as thirst is not quenched". From the representations mobilised to discursively legitimise this colonisation/rationalisation of space directed by the State, the idea of redemption will have a significant weight, as the commemorative inscription of the inauguration of the Cenajo reservoir on the Segura River perfectly illustrates:

> This Cenajo reservoir was ordered to be built by Francisco Franco, Caudillo of Spain. With it he controlled the turbulent waters of the Segura River so that they would peacefully fertilise fertile land and *redeemed* the men who work it from the millenary fear of floods and drought. With his presence it was inaugurated on 6 June 1963.

Therefore, the passive revolution, in terms of economic modernisation from above, promoted by Franco's dictatorship as a solution to the southern question in the southeast, was presented as an intervention on climatic irregularity and the solution to the "water deficit". Thus, state water management implied the institutionalisation of aridity as a collective threat from which it was a moral obligation to free oneself.

From the 1980s onwards, this moral race was to be led by the irrigator-farmer, as the promoter and creator of a new Eden – "the orchard of Europe" – as a counterforce to the advance of the desertification that always threatens the arid and semi-arid regions of the southeastern peninsular. This agricultural programme enjoys today enormous social consensus and support. For this reason, in the Region of Murcia, a broad coalition of interest groups will take up a collective position with the State to ensure that it periodically guarantees new water supplies (either through the Tajo-Segura water transfer or by promoting the construction of new transfers). This took shape a regional agenda of political mobilisation, progressively capitalised by the conservative forces, mainly articulated around the Partido Popular, which governed the Autonomous Community with very large absolute majorities from 1995 to 2015 and, in coalition with other forces, since 2015.

The fact that the solution to the southern question finally came from above, from the oligarchic forces, has had a significant influence on the monopoly of the agrarian agenda by the political right. At the same time, the resistance shown by the agrarian bloc to environmental groups' denunciations of the environmental impacts of intensive agriculture, and even to the environmental regulations and ecological transition policies promoted by the European Union, is an expression of the deep regional history. The narrative of the underdevelopment of the southeast due to lack of water and aridity is now being used as an argument against environmentalist groups who are accused of wanting to return to the situations of poverty from which they have worked so hard to emancipate themselves. It is no coincidence that with the exacerbation of the ecological crisis – climate change, restrictions on water resources, Mar Menor – the influence of far-right political forces has increased and, therefore, so have denialist positions.

Conclusion

The aim of this chapter has been to "understand" the agrarian rationale and the deep history of farmers in the Region of Murcia as a way to find a way out of the environmental conflict in the Mar Menor. To do so, we have followed a Gramscian perspective. According to Gramsci,

> in order to weave alliances it was necessary to make politics and for this it was necessary to investigate, to study, that is to say, to analyse the territorial and class composition of the country; it was necessary to examine the social classes themselves, which far from being homogeneous were (and are) composite and complex realities. The communists had to study the political tendencies, the culture, the conception of the world (including religion) of the classes against which it was necessary to fight, as well as those with which it was necessary to establish an alliance.
>
> (Tafalla, 2024, pp. 76–77)

In this sense, we advocate the need to articulate an alliance between political environmentalism and agrarianism to enable a just ecological transition of the European economic model that overcomes the authoritarian neoliberal project. The Mar Menor is a laboratory where these problems and potentialities can be glimpsed. For this political construction, it must be considered that the agrarian bloc is internally differentiated into different classes. This strategic alliance must be sought between the classes of labour, including small- and medium-sized farmers who occupy a contradictory class position (between the classes of capital and the classes of labour). For this articulation, a water policy proposal adapted to climate change and with an anti-oligarchic social sense, i.e. one that favours the working classes, will be fundamental.

Such an alliance must perhaps also adopt a methodological premise formulated by Piero Gobetti, a thinker of the Italian southern question of great influence on Gramsci: "not to adhere to one of the two formulas, but to understand the visions of the two opposing elements" (in Tafalla, 2024, p. 91).

Notes

1 Funding: This article has been possible thanks to the project AGROTRANSICIÓN funded by Ministry of Science, Innvovation and Universities of Spain (PID2023-150450NB-I00).
2 It is a term used by Gramsci to refer to social, political, and economic transformations induced from above by the ruling classes and in which the dominated classes do not actively intervene, although their aspirations are partially addressed to incorporate them into the process. For an application of the concept of passive revolution to Francoism, see Villacañas (2022).
3 Neil Smith (2020) differentiates between absolute space, as a universal receptacle that contains objects and in which events occur, and relative space, which is not independent of matter, and is in fact shaped by material relations.
4 It should be borne in mind that the law on the Tagus-Segura aqueduct guaranteed a contribution of a maximum of 540 hm^3 net. However, the average transfer of resources since the aqueduct came into operation is approximately 250 hm^3 (to date this average transfer has tended to decrease as a result of the prolonged droughts we have experienced).

References

ANSE & WWW-España (2018). *La burbuja del regadío: el caso del Mar Menor. Evolución de los regadíos en el entorno del Mar Menor. Campo de Cartagena (1977-2017).*

Baños, P., Pérez, I., & Pedreño, A. (2009). Aportaciones desde la investigación social al debate sobre agua y regadío. *Anduli: revista andaluza de ciencias sociales*, *8*, 83–97.

Costa, J. (1984.) *Oligarquía y caciquismo, Colectivismo Agrario y Otros Escritos*. Alianza Editorial.

Fanlo, A. (2002). La gestión del agua en España: experiencias pasadas, retos futuros. *Revista electrónica del Departamento de Derecho de la Universidad de La Rioja*, 0, 43–53.

Gramsci, A. (2024/1930). *La cuestión meridional*. Verso.

Hochschild, A. (2016). *Extraños en su propia tierra*. Capitán Swing.

Martínez Carrión, J. M. (2002): *Economía de la Región de Murcia*. Editora Regional de Murcia.

Martínez, J., & Esteve, M. A. (2000). Estimación de la entrada de nutrientes al Mar Menor utilizando un modelo dinámico. *Mediterránea*, *17*, 5–30.

Moore, J. W. (2020). *El Capitalismo en la trama de la vida.* Traficantes de Sueños.
Pedreño, A. (1999). *Del jornalero agrícola al obrero de las factorías vegetales.* Ministerio de Agricultura, Pesca y Alimentación.
Pedreño, A., & Riquelme, P. (2022). El trabajo asalariado agrícola en los territorios rurales españoles. Retos y oportunidades. *Mediterráneo Económico, 35*, 257–277.
Pérez Picazo, M. T., & Lemeunier, G. (1990). Agricultura y desarrollo regional en Murcia, 1750-1980. *Revista Áreas*, *12*, 225–236.
Pérez Picazo, M. T., & Lemeunier, G. (1994). La evolución de los regadíos mediterráneos. El caso de Murcia (siglos XVI-XIX). In Sánchez Picón, A. (ed.) *Agriculturas Mediterráneas y Mundo Campesino: Cambios históricos y retos actuales. Actas de las Jornadas de Historia Agraria de Almería, 19-23 de abril de 1993* (pp. 47–65). Instituto de Estudios Almerienses, Diputación de Almería.
Segura, P., & Pedreño, A. (2006). La hortofruticultura intensiva en la Región de Murcia, un modelo productivo diferenciado. In Etxezarreta, M. (coord.), *La agricultura española en la era de la globalización* (pp. 369–421). Ministerio de Agricultura, Pesca y Alimentación.
Smith, N. (2020). *Desarrollo desigual. Naturaleza, capital y la producción del espacio*, Traficantes de Sueños.
Tafalla, J. (2024). Introducción. In Gramsci, A. (ed.) (2024/1930), *La cuestión meridional* (pp. 9–136). Verso.
Villacañas, J. L. (2022). *La revolución pasiva de Franco.* Harper Collins.

4 The ecological degradation of the Mar Menor. Power, science and deep stories in Global agriculture[1]

Carlos de Castro Pericacho, Andrés Pedreño Cánovas and Miguel Ángel Sánchez García

Introduction

At an Assembly of the Sindicato Central de Regantes del Acueducto Tajo Segura (SCRATS) held in December 2017, Luis del Rivero, one of the most important figures in the agri-food sector in the Region of Murcia, said the following:

> For the first time we have a government in Madrid that is against the aqueduct, not for it as we have had before, and that changes things.
>
> For the first time in 37 years [since 1980] public opinion… is not in favour of the irrigators… we have to be aware that the battle of public opinion has been won for us… by the environmentalists… And we have to know that because we have to see how to reverse this issue.[2]

This intervention took place at a time of intense and historic social mobilisations in the Region of Murcia that pointed to the agricultural sector in the region as the main responsible for the discharge of nitrates and demanded legal and productive reforms to stop the environmental degradation of the lagoon.

In this sense, the eutrophic crisis of the Mar Menor in May 2016 meant the rupture of the social, media and political consensus on the environmental impact of intensive agriculture in the Region of Murcia. Since then, a real political and social battle for nature began, more specifically, for the explanation of the pollution and the search for solutions for the recovery of the Mar Menor.

As the quote that opens this chapter admits, in the early stages of this battle, farmers found themselves disoriented and discredited. The forcefulness of the images of a rotting lagoon, the emotional impact on the population, as well as the rapid and spontaneous mobilisation of the people overturned the hegemonic discourse of the agricultural sector.

Moreover, there was an important fracture in the hegemonic narrative of the heroic farmer, as it had been configured since the 1960s, according to which a new productive subject linked to irrigated agriculture was finally going to leave behind the secular underdevelopment of the peninsular southeast, by imposing itself over traditional rain-fed lands hit by drought (Pedreño et al., 2022; Sánchez-García et al., 2022).

DOI: 10.4324/9781003489078-5

This chapter aims to situate the regional political conflicts around agriculture and the Mar Menor in the context of a dispute over the appropriation of nature. The general idea is that all actors involved in this socio-environmental conflict are trying to appropriate nature culturally and politically in order to push for a kind of social change, either a just and emancipatory ecological transition or an ecological reconversion of the sector within the coordinates of green capitalism.

In these contested processes of appropriation of nature, actors, on the one hand, mobilise and produce different types of resources, values and knowledge and, on the other hand, directly or indirectly express different conceptions of the relationship between nature, economy and society.

This chapter will focus exclusively on how the agrarian coalition has mobilised the production of scientific knowledge in order to promote its own model of agrarian ecological transition. This coalition is made up of local producer companies, the main producer and exporter associations (PROEXPORT,[3] FECOAM,[4] COAG,[5] APOEXPA,[6] Fundación Ingenio, etc.), the irrigation communities grouped in the SCRATS,[7] the main commercial distribution chains and the regional administration.

This is intended to underline at least two points. First, companies and the entire network of associations linked to the sector must be seen as political actors beyond the productive sphere, insofar as the *sector's political activism* is aimed at bringing about structural changes in line with its conception of the ecological transition. Second, this corporate political activism is accompanied by something that can be called a *corporate epistemology*, which refers to the fact that business and market values (competition, productivity, efficiency, performance) operate as preferential criteria that guide the observation of reality and the production of scientific knowledge itself and, therefore, also guide the understanding of the relationships between nature, the economy and society as well as the very conception of the agrarian ecological transition.

On the other hand, the case of the Mar Menor must be placed in a broader context. On the one hand, the ecological disaster of the Mar Menor is not an isolated case, but another example of the enormous environmental impact of global agriculture (Friedmann, 2005; Sage, 2022). Numerous areas in Europe have been declared vulnerable to nitrate pollution, including the Campo de Cartagena since 2001. A 2018 European Commission report on the implementation of the nitrates directive between 2012 and 2015 showed that 19% of European rivers and 26% of European lakes were affected by eutrophication mainly from intensive agriculture and the livestock industry (European Commission, 2018). This means that the environmental problems of agriculture in the region are, to a large extent, problems stemming from the global agri-food system in which production and distribution are organised along global chains that are dominated by large food distribution groups and which connect, on the one hand, the places of intensive agricultural production (which bear a large part of the so-called negative externalities such as the increasing expulsion of small producers, extremely precarious working and living conditions of foreign workers, irreparable ecological damage) and, on the other hand, the places of consumption, i.e. the large retail outlets (Corrado et al., 2017).

On the other hand, the case of the Mar Menor can be presented as another example of the growing "climate unrest" on a global scale that is tearing apart the political and institutional order of our societies. The general debate concerns how this climate unrest is to be defined socially and politically. Measures to mitigate the effects of climate change force the questioning of at least two of the premises on which the political and economic institutions of our societies, as well as their class structure, have been built: the use of fossil fuels and the ideology of growth. Therefore, this type of socio-environmental conflict opens the opportunity to debate how the relations between nature and society can be rearticulated, without there being any predetermined political meaning to this ecological transition: whether a more democratic or authoritarian political and social order, or a new eco-reactionary order, will end up taking shape. The ecological transition opens a new scenario of political opportunities. And in each territory, it will be articulated in different ways according to its political and cultural history and according to the capacity of the social and political actors to mobilise resources to achieve their model of ecological transition. In this sense, the Region of Murcia, and specifically the socio-environmental conflict of the Mar Menor, represents one more stage in this global dispute to define a new social, economic and political order in which nature will play a decisive role.

The rest of the chapter is organised as follows. First, it will indicate what it means that the socio-environmental conflict around the Mar Menor is a dispute over Nature. Second, drawing on the contributions of Charles Tilly (1978) and Sydney Tarrow (1997), it will identify the types of resources that actors mobilise in their collective actions driven by their aspirations to produce a specific type of ecological transition. And third, drawing on what Naomi Oreskes and Eric Comway (2010) have called the tobacco strategy, this chapter will describe how the agrarian coalition has used the production of scientific knowledge in its resource mobilisation strategy to engage in the dispute over nature.

Disputing nature

The socio-environmental conflict of the Mar Menor has placed nature at the centre of the political debate and has become the axis that is restructuring political and social alliances throughout the Region. However, nature and territory have occupied an important place in the region for decades. At least since 1995, after the previous severe industrial restructuring (Ibarra, 2016), the Region's development strategy has been based on three sectors, construction/urbanism, tourism and agriculture, which intensively consume natural resources and are very aggressive towards the territory.

This strategy is underpinned by a conception of nature as an external, abstract entity, separate from society that can provide unlimited resources. Nature appears as something to be conquered or mastered (Swyngedouw, 2015).

In the case of agriculture, this conception of nature appears in what can be called the *myth of the heroic farmer,* which underlies the promotion of the Region as the Orchard of Europe (in national media, international fairs). This myth began to be forged more strongly at the end of the 1990s and praises a generation of farmers and irrigators who from the late 1960s took part in a collective enterprise that turned a semi-arid land into the market garden of Europe, into one of the most productive, most profitable, and most biotechnologically avant-garde agricultural territories in global agriculture. An irrigated agriculture that, at last, was to leave behind the secular underdevelopment of the southeast peninsular, imposing itself over the traditional drylands scourged by drought and scarcity.

Probably, the "Agua para todos" (Water for all) campaign, promoted since the early 2000s by the government of the Murcia region on behalf of the entire agricultural sector to demand from the state government a new water transfer from the Ebro River to the Segura river, was the high point in the constitution of this myth and, in general, managed to place agriculture and water demands at the centre of regional political life, while shaping a strong regional identity, reinforcing political, business and social alliances and garnering overwhelming social support.[8]

The eutrophication crisis of May 2016 shatters this myth and opens a dispute over the political and cultural appropriation of nature, a dispute to forge a new link between nature and agriculture.

Social mobilisations were able to turn a problem of environmental degradation into a cultural and political crisis in the agricultural sector. Civil society associations (mainly Pacto por el Mar Menor,[9] made up of a wide and diverse network of associations including environmental groups such as ANSE,[10] WWF[11] and Ecologists in Action, neighbourhood associations, fishermen's associations, etc.) deployed a discourse that took up what numerous scientific bodies had been warning about for a long time and which emphasised the link between the pollution of the Mar Menor and intensive agricultural practices (Martínez and Esteve, 2000). However, it was the first time that the sector had been held politically and socially responsible in such a massive way. It was a discourse that circulated through assemblies, rallies, demonstrations, and the media and found unprecedented social support.

The demands of organised civil society aimed to establish a different relationship with nature, prioritising the conservation of biophysical balances and questioning whether intensive agriculture could respect these balances, which is why important productive reforms of the sector were demanded, such as a reduction in the use of nitrates and water, as well as a reduction in irrigated land, among others.

In these demands, a conception of nature becomes palpable that is more like the one envisaged by sociologists such as Jason Moore (2020) or Erik Swyngedouw (2015), according to which there is no separation between nature and society, but rather an assumption that the two are intertwined without it being possible to see where one begins and the other begins. There is thus a relationship of mutual interdependence between the human and non-human elements that make up the diffuse nature-society complex that changes over time.

Resources for change

What are the resources that actors mobilise in this dispute over nature and what changes do they aspire to bring about? Here, this chapter will loosely follow Charles Tilly (1978) and Sydney Tarrow (1997), who are two of the sociologists who have best studied the emergence of collective action phenomena and their eventual contribution to social and political change.

According to these authors, three phases can be distinguished in collective action phenomena. On the one hand, we have the structure of political opportunities, which refers to the social, political, cultural and, we can add, environmental conditions in which new political subjects can come to be constituted and which can eventually drive some collective actions. On the other hand, we have the repertoires of collective action, which refer to the organisational, political, economic and cultural resources that collective subjects can use in their own configuration process, on the one hand, and, on the other, in the imagination, design, planning and implementation of collective actions, as well as in the imagination of the social change they want to produce. And finally, we have the outcomes of collective actions, which are not predetermined by the conditions, or by the structure of the conflict situations. Thus, the outcome of these collective actions can be emancipatory or reactionary, depending on the correlation of social forces and the resources employed by the actors. In other words, the outcome is contested, open and contingent. In other words, history is neither written, nor does it follow an inexorable course towards some end, nor is it on our side or against us.

From this point of view, the collective actions of social and neighbourhood movements can be analysed, but also those of the agrarian coalition. In other words, it is possible to analyse the agricultural companies and the associative fabric around them as social and political activists.

Another important issue here is to look at how actors imagine agrarian ecological transition. The available resources can also guide this imagination of agrarian social change or agrarian ecological transition.

The solutions proposed by the agrarian coalition fit into the *technomarket* approaches to environmental problems (Levy and Spicer, 2013) insofar as it proposes to address these problems through technological innovations carried out in the market by entrepreneurs. In this approach, it is not regulation, but rather the dynamism of the market itself that will find the most optimal solution and it will be private technological innovation that will optimise the use of resources.

Merchants of Doubt in Murcia

This chapter will focus on part of the repertoires of collective action, i.e. part of the economic, social, cultural, organisational and cognitive resources mobilised by the actors. More specifically, it will focus on describing how the agrarian coalition has mobilised the last two, organisational and cognitive resources (science) to participate in the dispute over the appropriation of nature or, as indicated in the quote at the beginning, to "…see how to reverse this issue".

In order to try to interpret the resource mobilisation strategy followed by the agrarian coalition during this time, it is proposed to resort to what Oreskes and Conway (2010) have called the "tobacco strategy" in their book *The Merchants of Doubt*.

This strategy refers to how the tobacco industry, when the first clear scientific evidence of the link between smoking and cancer and other cardiovascular diseases was discovered in the 1960s, began to fund an infrastructure of ostensibly independent think tanks led by eminent and renowned scientists. Oreskes and Conway (2010) distinguish four phases in this strategy, which may overlap in time.

1 *Sowing doubt.* Industry challenged the scientific consensus either by doubting the available evidence or by pointing to other possible causes.
2 *Diverting attention.* Faced with the impossibility of unequivocally determining a single factor as a cause and the relative importance of each potential causal factor, the industry demanded and/or funded further studies investigating these other potential causes to remove uncertainty.
3 *Warn of the eventual catastrophe.* Industry denounced that without full certainty, it is not possible to justify actions that are very costly and that, in addition, can destroy jobs and break the economy.
4 *Unveiling the hidden agenda.* And if there is no absolute certainty, which is impossible in all science, and there is insistence on taking action against the industry, then there must be some hidden agenda that needs to be uncovered. So, the industry tried to denounce that the proposed actions, usually large-scale regulations by the state, would lead to undermining individual freedom by influencing the habits and lifestyles of the population and imposing a new lifestyle.

Subsequently, other industries found it to be an effective strategy and adopted it to address various environmental problems (mining, pharmaceuticals) such as acid rain, asbestos diseases, the impact of greenhouse gases on the ozone layer or CO_2 emissions as a cause of climate change. This has only delayed the implementation of measures needed to address the problems.

With a great deal of caution, and with some distance, this chapter will try to use this analytical framework to interpret the resource mobilisation strategy promoted by the agrarian coalition to deal with the case of the Mar Menor.

Sowing doubt

Prior to 2016, there was already a broad scientific consensus on the origin of the pollution of the lagoon, which began to consolidate with a study carried out before the year 2000 that warned of the growing impact of irrigated agriculture on the Mar Menor (Martínez and Esteve, 2000). In a more recent paper, Julia Martínez (2022), then director of the New Water Culture Foundation, reviewed a large sample of studies from various institutions such as the Spanish Institute of Oceanography and several universities carried out by active scientists and with recent fieldwork since 2016, and even earlier, which demonstrated

> that the Mar Menor is eutrophicated; that it is eutrophication that has caused the degradation of the lagoon; that the main source of nutrients comes from intensive agricultural activities (agriculture and livestock) in the Campo de Cartagena; that, in the case of nitrogen, the main source is agricultural fertilisers from crops; and, finally, that nutrient flows enter mainly via the surface and, therefore, are produced by current agricultural activities and

> not so much by pollution caused by past activities and accumulated in the Quaternary aquifer.
>
> (Martínez, 2022, p. 2)

These were conclusive studies which, nevertheless, did not fail to recognise the influence, albeit minor, of other factors such as filtrations from the quaternary aquifer with its water contaminated by past and present agricultural practices, wastewater from undersized and poorly maintained sewage treatment plants in some riverside municipalities, boats or filtrations from mining ponds.

However, from the very moment that the "green soup" appeared in May 2016, the debate about the other causes began to take centre stage. For example, according to information from critical members of the Scientific Committee,[12] the regional government allegedly ignored or questioned the available scientific evidence, demanded and pushed for more studies on the situation, and disseminated diagnoses and implemented actions that either diverted attention from agriculture or included it in a tokenistic way.

The scientific debate on the causes of the pollution of the Mar Menor came to be expressed in terms of how to quantify the weight of each factor and about the description of the development of the impact of each factor in order to take the most effective measures. This debate fractured the Committee. It was a committee in which scientists were in a minority compared to a large number of technicians and officials from the different directorates-general of the ministries.[13] The institutional positions and some scientists were dragging the committee towards factors other than agriculture. Meanwhile, another group of scientists critical of the official version tried to focus attention on agriculture, without excluding the importance of other factors. This tension led to the resignation of four members of the Committee by June 2018 (Rosa Gómez, Francisca Giménez Casalduero, Julia Martínez and Miguel Ángel Esteve).[14] In October 2019 another four scientists also resigned (Gonzalo González, Juan Manuel Ruiz, José Alvárez Rojel &Víctor León).[15] The reasons in both cases were the same as in the letter of protest signed by 11 scientists of the committee a few months earlier[16]: instrumentalisation by the regional government, lack of transparency in the criteria of composition and functions, and the political, fragmented and partial use of the scientific reports both to determine the causes and to justify actions to repair the lagoon.

The result was the spread of doubt and uncertainty and a shift of attention away from the proven main cause, nitrogen fertilisers from intensive agriculture to other factors that were undoubtedly also involved (sewage treatment plants in poor condition, boats, seepage from mining ponds), but to a lesser extent.

Diverting attention

This shift in the debate led to the demand for, and even the funding of, new studies. And this is a moment in which the very production of scientific knowledge is going to become an object of dispute. Since about 2017, a whole network of "chairs" ("Cátedras"), which are agreements between public universities and private organisations, started to be created. The agreements between several public universities in the region with various agricultural organisations were aimed at creating their own scientific evidence that either focused their attention on these other factors of lesser weight or questioned the scientific evidence consolidated by dozens of previous studies.

For example, in 2017, the Sindicato Central de Regantes del Acueducto Tajo Segura (SCRATS), Comunidad de Regantes de Campo de Cartagena (CRCC) together with the Universidad Politécnica de Cartagena (UPCT) created the Cátedra de Trasvase y Sostenibilidad. The coordinator of the Chair, Mariano Soto, indicated in an interview that "our task is to seek solutions in a world that is now highly mediatised, but which lacks rigorous *technical analysis*".[17]

Another important example is the Chair of Sustainable Agriculture of Campo de Cartagena, promoted by the Federation of Agricultural Cooperatives of Murcia (FECOAM) and the Coordination of Farmers and Stockbreeders Organisations (COAG) together with the UPCT. One of its main lines of work is research into the denitrification of brine using wood bioreactors.[18] On this scientific basis, the regional government has promoted since the end of 2021 a project for the construction of 15 wooden basins in Los Alcázares with the capacity to denitrify 6.3 hm^3/year.[19]

Projects like this can certainly be part of the solution, but they shift attention and resources away from the obvious problem of increasing nitrate application for more than 20 years.

According to statistics on fertiliser consumption in agriculture, compiled by the Ministry of Agriculture, some 800,000 tonnes of nitrogen fertilisers were used in the Region of Murcia between 2005 and 2020, an annual average of almost 50,000 tonnes. Logically, however, not all this amount was applied in the Campo de Cartagena, nor did it end up in the Mar Menor. A 2017 report, commissioned by the Ministry of Ecological Transition and carried out by a committee of experts, estimated that the contribution of the agricultural activity carried out in the area to the nitrate content of the aquifer was around 3,300 tonnes/year (Mar Menor Scientific Advisory Committee, 2017, p. 18), of which a significant part ends up in the lagoon. At the same time, if one considers that the approximately 45,000 ha of irrigated cropland in Campo de Cartagena need, according to the estimate of one of the CRCC managers, around 150 hm^3 of irrigation water per year, it can be concluded that the scope of the bioreactors is limited.

Another example of the production of scientific knowledge as an object of dispute was the study commissioned by the Comunidad de Regantes del Arco Sur also in 2017 to try to quantify quaternary discharge. The report of this study estimated a potential discharge of 71 hm^3/year and an actual discharge of between 38 and 46 hm^3/year (Contreras et al., 2017). However, the Ministry of Ecological Transition (MITECO) subsequently provided very different figures through the results of a report it commissioned from Tragsa (2018), which indicated that underground discharge was between 8.5 and 11.6 hm^3/year, well below the figures provided by the study of the Comunidad de Regantes del Arco Sur.

Once this series of studies is publicly available, it can be used, following the "tobacco strategy", to scientifically justify a programme of actions that prioritises factors other than agriculture and where the reduction of nitrogen fertiliser application is secondary.

Warn of potential catastrophe or risk and unveil the true objectives of the adversaries

All the business associations, irrigation communities and agricultural associations in the sector (PROEXPORT, FECOAM, SERATS, COAG, etc.) have played a decisive role in these two phases. However, the role played by the Ingenio Foundation (Fundación Ingenio) has marked a decisive milestone in this dispute for Nature. The Foundation was created in May 2020 and its board of trustees includes numerous associations and companies from the sector.[20]

Their intervention in this dispute over the appropriation of nature has helped to give coherence to a discourse that was already circulating among producers but was not clearly reaching society.

The Ingenio Foundation has drawn up a plan, "Anillo Protector Ambiental" (Environmental Protection Ring),[21] with the aim of providing "an integral and definitive solution to recover and protect the Mar Menor".[22] It is the most complete and elaborated plan among the actors of the agricultural coalition and coincides to a large extent with the regional government's plan.

The campaign to publicise the "Environmental Protection Ring" was intense from the outset. They have occupied an enormous amount of space in the local and national media, organised events to make visible their commitment to the region's agricultural tradition and to nature, as well as to promote cutting-edge technology and precision agriculture. They have been received by representatives of the Ministry of Agriculture and European Union institutions, they have organised numerous publicity campaigns, they have created very effective slogans ("The solution is not to destroy agriculture"[23]) and they have named their main adversary, "blackmailing environmentalism", those who "before we arrived nobody stood up to them or challenged them".[24]

On the other hand, the Ingenio Foundation has been the main spokesperson for the sector in expressing its rejection of the environmental protection measures of the State and the EU, which it considers a threat to the sector. These measures include halving the use of pesticides and fertilisers and increasing by 25% the areas dedicated to organic farming, as well as various environmental regulations advocated by some environmental groups, the Ministry of Ecological Transition and the institutions of the European Union and included in the EU's Farm to Table Strategy, which aims for an ecological transition of the entire food system, trying to achieve carbon neutrality and reduce the loss of biodiversity by 2050.

Some public interventions by the CEO of the Ingenio Foundation indicate that this perception of the threat to the sector and to the way of life it supposedly embodies. For example:

> a group of farmers… who want to maintain Murcian agriculture as a national and European benchmark in terms of sustainable and innovative production…

> Farmers who have prospered thanks to the efforts of generations of entire families... .... But this legacy is now under threat.[25]
>
> (Natalia Corbalán, CEO of Fundación Ingenio, Appearance in the Assembly of the Region)

> ... There is intentionality in blaming the agricultural sector and other interests are masked or certain groups simply take advantage of the artificial and destructive image that has been created of the sector... Farmers are the victims of the [pollution] problem.[26]
>
> (Natalia Corbalán)

Results of collective actions

One of the main results was to spread doubt among the actors in the sector about the scientific knowledge available. Perhaps, the assessments made by some representatives of agricultural associations in interviews carried out in 2021 about the Law on urgent measures to guarantee environmental sustainability in the Mar Menor approved in 2018 and the questioning of its technical and scientific nature can serve as an example:

> One of the problems with the Mar Menor Law is that it is not based on technical criteria and contradicts common sense.
>
> (Head of the agricultural organisation, UPA)

> The law has neither head nor tail because it is technically unsupportable.
>
> (Head of irrigation community, CRCC)

> The best way to reconcile agriculture and sustainability is to legislate with scientific criteria, which have not been taken into account. It is legislating on a whim. It [the current Mar Menor law] seems to me to be a total and absolute nonsense.
>
> (Head of the agricultural organisation, COAG)

On the other hand, another result of this strategy has been to shift the population's attention from the Mar Menor to water. That is, from concern about environmental degradation to concern about the availability of water for the sector.

According to the latest CEMOP (Centro de Estudios Murcianos de Opinión Pública) barometer of winter 2023, concern about water was the biggest problem for 14.8% of the citizens of the Region, only behind unemployment, while the situation of the Mar Menor only concerned 4.4% of the population. In autumn 2019, it was the situation of the Mar Menor that most concerned the population (15.7%), only behind employment, and concern for water did not even appear in the barometer.

Why? Chains and deep stories

So far, it has been shown how the agrarian coalition has participated in the dispute over nature by deploying a network of alliances (chairs) and promoting the production of a type of scientific knowledge oriented towards its conception of the ecological transition in the sector.

Now, to conclude, this chapter will explore the why, the possible causes. Why are environmental protection measures and scientific evidence disputed, and why are demands for environmental protection and ecological reforms of agricultural production seen as a threat by the sector?

Position in the global agri-food chain

The structurally subordinate position that local producers occupy in the global agri-food chain can be considered as one of the main indirect causes of the profound ecological impact on the Mar Menor.

From this position, it is difficult for local producers to challenge supermarkets for greater involvement in solving environmental problems. By not questioning inequality, a techno-commercial solution is activated that tries to mitigate or displace the problem in order to keep the power structure intact and, therefore, the mechanisms of value distribution, as well as responsibility for the ecological impacts of production. We refer here to mechanisms or strategies such as those of productive differentiation, which have had environmentally pernicious effects but have been successful in competitive terms for agricultural companies in the region, as well as the increase in the area of irrigated land cultivated to take advantage of the benefits of scaling up production, increased productivity of soils with less fertile organic matter that require greater amounts of fertilisers and water, or the cultivation of more productive non-native varieties and the reduction of varieties due to the selection of the most commercially "suitable" ones that reduce biodiversity (de Castro et al., 2017).

Together with labour management strategies, these production strategies allow the sector to maintain a high competitive position in the chain since, on the one hand, they increase profit margins by optimising costs due to the impossibility of increasing prices dictated by supermarkets and, on the other hand, these strategies are profitable because they allow shifting or outsourcing the responsibility for managing environmental consequences to other agents, either to consumers or to public institutions (de Castro et al., 2017).

The point is that the incorporation of more moderate ecological demands into agricultural production processes and the management of their effects, included in public regulations, would weaken their competitive position in the global value chain. Hence, they are more inclined to accept private regulations or supermarket standards (de Castro et al., 2021), since compliance with them guarantees access to international markets, and less inclined to accept international, state or local public regulations, which would undoubtedly improve the situation of the Mar Menor but would make it more difficult for them to access markets by increasing their operating costs. And this is the dilemma in which local producers are trapped in Murcia and in all the productive territories of intensive agriculture in the global agri-food system. In other words, within a chain dominated by supermarkets, they cannot be non-polluting. Or, as Marina Requena summarises, to be competitive, you have to be polluting. Therefore, any proposal must take into consideration this subordinate position in the chain and challenge supermarkets and other intermediaries as part of the solution.

Loss of place in the world

However, it is not only the unequal position in the chain that explains the farming coalition's hostile response to demands for regulation and environmental protection. The concern that ecological transition programmes might wipe farming off the map expresses a deeper fear that with it, with agriculture, a way of life and a socio-entrepreneurial and regional identity that, according to their account, generation after generation of farmers has been consolidated in the region, will also be wiped out.

The presentation page of the Ingenio Foundation states that

> we defend an agriculture committed to its people and the rural environment… We are rooted in the land, committed to its people and to a rural environment that we defend *as an ancestral and sustainable way of life*, for today's families and future generations.

Recently, at the presentation of a Conference on Agriculture and Water in the Levante region in September 2022, the president of the Sindicato de Regantes del Trasvase Tajo Segura, Lucas Jiménez, indicated that the debate aims to shed light "on the future of our irrigation systems, of the aqueduct, of *our way of life*, in short".[27]

This fear of the "loss of our way of life" through the interference of actors who are situated in a threatening exteriority (environmentalists, the state government, the EU) and who are therefore not part of "us", is part of a logic of constructing emotional identifications that transcend supposed objective interests and aim at what Arlie Hochschild (2016) calls "a deep story", that is "a story of what one/one group feels", "a story in which feelings are told using a language of symbols and eliminating (or at least attenuating) the (supposedly) rational".

She studied the state of Louisiana to understand why its inhabitants, dramatically affected for decades by pollution from the oil industry, were massively supportive of political options, such as the Tea Party, an ultra-conservative section of the Republican Party, which advocated the elimination of all environmental protection, something that in principle ran counter to their "objective interests". In his interviews, Hoschchild found a deep history according to which "oil symbolised restored honour" in the face of a past of "plantations, slaves and poor tenant farmers" that brought "shame" to the South. In this "deep history", the environmental demands of environmentalists and the federal government represented a threat to this restored honour, to this pride in having escaped the past of plantations and slaves.

Is there a deep history of the farmers and irrigators of Campo de Cartagena? Probably yes, probably that history is made up of the shame of a past of poverty and emigration linked to traditional dry farming and a present in which, with their own efforts, they have converted a semi-desert land into one of the largest irrigated orchards in Europe of which they are proud. It is this diffuse fear of the loss of restored honour that may help us to explain much of the configuration of an environmental sensibility that moves the farming coalition away from regulation-based solutions and towards technology- and market-based solutions to preserve the region's intensive agriculture.

Hence, the creation of a broadly shared environmental sensibility that facilitates a just ecological transition must take into account not only the structural position of local producers, but also that deep history, thus exploring a space of collaboration and complicity between environmental demands and the demands of local producers.

Notes

1 Funding: This article has been possible thanks to the project AGROTRANSICIÓN funded by Ministry of Science, Innvovation and Universities of Spain (PID2023-150450NB-I00).
2 La Verdad, 25 January 2018. https://www.laverdad.es/murcia/rivero-gobierno-madrid-20180124235216-nt.html.
3 Association of Fruit and Vegetable Producers-Exporters of the Murcia Region. (Asociación de Productores-Exportadores de Frutas y Hortalizas de la Región de Murcia)
4 Federation of Agricultural Cooperatives of Murcia. (Federación de Cooperativas Agrícolas de Murcia)
5 Coordinator of Farmers' and Stockbreeders' Organisations of Murcia. (Coordinadora de Organizaciones de Agricultores y Ganaderos de Murcia)
6 Association of Producers – Exporters of Fruit, Table Grapes and Other Agricultural Products. (Asociación de Productores – Exportadores de Frutas, Uva de Mesa y Otros Productos Agrarios)
7 Central Irrigation Union of the Tagus-Segura Aqueduct. (Sindicato Central de Regantes del Acueducto Tajo-Segura)
8 https://www.elsaltodiario.com/camara-civica/agua-para-todos-historia-de-la-campana-que-explica-por-que-en-murcia-siempre-gana-el-pp.
9 Pacto por el Mar Menor is a platform made up of various associations and individuals that was set up with the aim of making the defence of the Mar Menor more visible, https://pactoporelmarmenor.blogspot.com/p/quienes-somos.html.
10 Asociación de Naturalistas del Sureste (Southeast Naturalists Association).
11 Worldwide Fund for Nature.
12 *La Verdad*, 17 May 2018. https://www.laverdad.es/lospiesenlatierra/divorcio-comite-cientifico-20180513015143-ntvo.html.
13 *La Verdad*, 17 May 2018. https://www.laverdad.es/lospiesenlatierra/divorcio-comite-cientifico-20180513015143-ntvo.html.
14 *La Verdad*, 17 May 2018. https://www.laverdad.es/lospiesenlatierra/divorcio-comite-cientifico-20180513015143-ntvo.html.
15 *La Verdad*, 22 October 2019. https://www.laverdad.es/lospiesenlatierra/noticias/cuatro-expertos-abandonan-20191020213140-nt.html.

16 *La Verdad*, 5 September 2019. https://www.laverdad.es/lospiesenlatierra/noticias/quince-cientificos-rechazan-20190905095430-nt.html.
17 *La Verdad*, 26 March 2019. https://www.laverdad.es/agro/contaminacion-politica-tapa-20190326011222-ntvo.html.
18 https://www.catedraagriculturasostenible.es/2021/07/26/la-catedra-de-agricultura-sostenible-y-golftat-prueban-un-sistema-para-desnitrificacion-de-salmueras-altamente-concentradas/v.
19 *El Español*, 4 October 2021. https://www.elespanol.com/ciencia/medio-ambiente/20211004/biorreactores-astillas-gobierno-murcia-nitratos-mar-menor/616939031_0.html.
20 https://fundacioningenio.com/.
21 https://fundacioningenio.com/solucion-mar-menor/.
22 https://fundacioningenio.com/wp-content/uploads/2021/12/DIPTICO-ANILLO-PROTECTOR-AMBIENTAL.pdf.
23 https://fundacioningenio.com/noticias/.
24 Natalia Corbalán's speech at the celebration of the Foundation's first anniversary. https://fundacioningenio.com/la-fundacion-ingenio-cumple-su-primer-ano/
25 https://hermes.asambleamurcia.es/documentos/pdfs/ds/DS_10/COMISION/ESPECIALES/MAR%20MENOR/CEMARME210301.008.pdf
26 https://hermes.asambleamurcia.es/documentos/pdfs/ds/DS_10/COMISION/ESPECIALES/MAR%20MENOR/CEMARME210301.008.pdf
27 La Verdad, 25 September 2022. https://www.laverdad.es/eventos/lideres-politicos-expertos-20220925205417-nt.html?fbclid=IwAR32UqkvlJPgBXqcuA9IKWaknQ56Xov3XeDOISFSoaXcPBADCs51r2k7a_0.

References

Contreras, S., Alcolea, A., Jiménez-Martínez, J., & Hunink, J. E. (2017). Quantification of groundwater discharge to the Mar Menor by hydrogeological modelling of the Quaternary surficial aquifer. Available at: https://www.futurewater.es/projects/descarga-mar-menor/. (Last accessed March 30, 2024).

Corrado, A., de Castro, C., & Perrotta, D. (2017). Cheap food, cheap labour, high profits. Agriculture and mobility in the Mediterranean. In Corrado, A., de Castro, C., & Perrotta, D. (eds.), *Migration and Agriculture: Mobility and Change in the Mediterranean Area*. (pp. 79–94). Routledge.

de Castro, C., Gadea, E., Pedreño, A., & Ramírez, A. (2017). Coaliciones sociales y políticas en el desarrollo del sector agroexportador: las frutas murcianas en las redes globales de producción agroalimentaria. *Mundo Agrario, 18*(37), e043. https://doi.org/10.24215/15155994e043.

de Castro, C., Gadea, E., & Sánchez-García, M. A. (2021). Estandarizadores. La nueva burocracia privada que controla la calidad y la seguridad alimentaria en las cadenas globales agrícolas. *Revista Española de Sociología, 30*(1), a16. https://doi.org/10.22325/fes/res.2021.16.

European Commission (2018). *Report from the Commission to the Council and the European Parliament on the Implementation of Council Directive 91/676/EEC Concerning the Protection of Waters against Pollution Caused by Nitrates from Agricultural Sources Based on Member State Reports for the Period 2012-2015*. Available at: https://ec.europa.eu/environment/water/waternitrates/pdf/nitrates_directive_implementation_report.pdf (Last accessed March 30, 2024).

Friedmann, H. (2005). From colonialism to green capitalism: Social movements and the emergence of food regimes. In Buttel, F. H., & MacMichael, P. (eds.), *New Directions in*

the Sociology of International Development. Research in Rural Sociology and Development 11 (pp. 227–364). Elsevier.
Hochschild, A. (2016). *Extraños en su propia tierra.* Capitán Swing.
Ibarra, J. (2016). *Cartagena en Llamas: la crisis industrial de 1992.* Ediciones Corbalán.
Levy, D. L., & Spicer, A. (2013). Contested imaginaries and the cultural political economy of climate change. *Organization, 20*(5), 659–678.
Mar Menor Scientific Advisory Committee (2017). *Comprehensive Report on the Ecological Status of the Mar Menor.* Available at: https://www.europarl.europa.eu/cmsdata/245205/BRIEFING.pdf (Last accessed March 30, 2024).
Martínez, J., & Esteve, M. A. (2000). Estimación de la entrada de nutrientes al Mar Menor utilizando un modelo dinámico. *Mediterránea, 17*, 5–30.
Martínez, J. (2022). *El Mar Menor. Falacias y realidades.* Fundación Nueva Cultura del Agua.
Moore, J. W. (2020). *El Capitalismo en la trama de la vida.* Traficantes de Sueños.
Oreskes, N., & Conway, E. (2010). *Merchants of Doubt. How a Handful of Scientists Obscured the Truth on Issues from Tobacco Smoke to Global Warming.* Bloomsbury Press.
Pedreño, A., de Castro, C., & Sánchez-Rodríguez, M. A. (2022). Producir la naturaleza: Agricultura intensiva, estándares de calidad y controversias ambientales en el Mar Menor. In de Castro, C., Reigada, A., & Gadea, E. (eds.), *La producción de la calidad en el sector agroalimentario: un análisis sociológico* (pp. 17–78). Tirant Lo Blanch.
Sage, C. (ed.) (2022). *A Research Agenda for Food Systems.* Edward Elgar Publishing.
Sánchez García, M. Á., Pedreño, A., & de Castro, C. (2022). The nature of standards: How standards shape the value of nature. *International Sociology, 37*(6), 612–629. https://doi.org/10.1177/02685809221115962.
Swyngedouw, E. (2015). *Liquid Power: Contested Hydro-Modernities in Twentieth Century Spain.* MIT Press.
Tragsa (2018). *Cuantificación, control de la calidad y seguimiento piezométrico de la descarga de agua subterránea del acuífero cuaternario del Campo de Cartagena* (CLAVE: 07.831-0070/0411). Ministerio de Ciencia, Innovación y Universidades. Instituto Geológico y Minero de España.
Tarrow, S. (1997). *El poder en movimiento: Los movimientos sociales, la acción colectiva y la política.* Alianza Editorial.
Tilly, C. (1978). *From Mobilization to Revolution.* Random House McGraw-Hill Publishing.

5 Impacts of mining in the Sierra Minera on the Mar Menor lagoon

Pedro Baños Páez and Isabel Banos-González

Introduction

The context of Sierra Minera

The mining district of Cartagena-La Unión, called Sierra Minera, with a surface area of approximately 114 km^2, stretches for about 25 km in an East-West direction, from Cabo de Palos to Cartagena, in the South-East of the Iberian Peninsula (Figure 5.1). The Sierra Minera is distributed over the municipalities of Cartagena and La Unión and it discharges its runoff waters into the Mediterranean Sea to the South and into the Mar Menor to the North.

The geological structure of the Sierra Minera is characterised by thrust layers with superimposition of different tectonic-stratigraphic complexes. This tectonic complexity made up of mantles, seams, disseminations and seams, together with various episodes of volcanism and other hydrothermal phenomena, have resulted in an extraordinary geology, with many and very diverse mineralisations that have been the object of mining workings during different periods. Recent discoveries show evidence of copper, lead and silver ore concentration works around 2,500 years B.C. in the vicinity of the current port of the city of Cartagena.

After the intense mining carried out by the Phoenicians, Carthaginians and Romans until the 3rd century A.D., a period of lethargy began until the mid-19th century. From the discovery in 1839 of the Jaroso vein in the Sierra de Almagrera, in the neighbouring province of Almería, a real mining fever broke out, reactivating the mining activity in the Sierra Minera of Cartagena-La Unión. This intense period of mining boom came to an end at the beginning of the 20th century, after the end of the First World War.

In the early 1930s, the Zapata-Portmán Mining and Metallurgical Company (SMMZP) was set up, with the French company Société Minière et Métallurgique Peñarroya and the Mancomunidad Zapata-Portmán, each owning 50% of the shares. In 1940, the SMMZP, after some small pilot experiments, started up a washery for the concentration of minerals in the Regente-El Concilio mining area, placed in El Gorguel (Cartagena), incorporating a new technique of differential flotation as an ore concentration system. This new technology, imported from Australia, where it had been developed since the end of the 19th century, represented a real revolution

DOI: 10.4324/9781003489078-6

in our country in the mineral concentration system, differing from the traditional gravimetric concentration methods. Differential flotation systems made it possible to exploit more diverse and less mineral-rich ores. They also required large quantities of chemical compounds, such as sodium cyanide, sulphuric acid, potassium xanthate and copper sulphate, depending on the different minerals being exploited.

In 1957, with full control of the Mancomunidad Zapata-Portmán by the SMM Peñarroya, the "Roberto washing plant" was commissioned next to the bay of Portmán, in Sierra Minera. It was presented as one of the largest installations in Europe for the concentration of minerals by the differential flotation system. Its initial treatment capacity was around 1,000 t/day, which was gradually increased to 10,000 t/day in the final stage of operation. This infrastructure is complemented by the entry into service of the mining train that ran from the Secondary Crushing, next to the Roberto washing plant, to the centre of the Sierra Minera with a route of more than 3 km, transporting the mining material from the various mines to the Roberto washing plant. Also noteworthy in this period controlled by the SMMZP is the gradual replacement of underground mining by open-cast mining, causing gigantic earthworks to access the mining deposits, resulting in some 15 km^2 of the Sierra Minera deeply transformed by these open-cast mining operations.

This prolonged mining activity, particularly intense and intensive in the last half of the 20th century, leaves, as aforementioned, a bleak picture in the Sierra Minera of Cartagena-La Unión and on the adjacent coastal edges, the Mar Menor to the North and the Mediterranean Sea to the South.

Several authors (Conesa et al., 2006; Belmonte Serrato et al., 2010; Romero Díaz et al., 2011; Banos-González et al., 2013, 2017; Baños-Páez, 2023) have described the changes suffered in the landscape of the Sierra Minera where the extractive activities took place. The enormous volumes of extracted material have configured a new morphology of the Sierra Minera, in a kind of telluric landscape, where the mountains move, creating enormous "vacies" or "terreras" in which the waste earth from the quarries that could not be treated was deposited; as well as the old mining tailings swamps ("pantanos") of the small differential flotation laundries of the Sierra Minera that did not discharge into the sea, as described below.

Socio-environmental impacts of the mining activities

The existence of old mining structures and remains gives rise to a number of notable socio-environmental impacts and health risks (Levasseur et al., 2022). As highlighted by Belmonte Serrato et al. (2010), it has been estimated that the number of deposits in Sierra Minera is 2,351. The ones that occupy the largest surface area are the 89 "flotation sludge ponds" or swamps that drain directly into ephemeral channels, as well as the 32 "vacies" or "terreras". The waste has no containment dykes, so there is a continuous direct contribution to the ephemeral channels through water erosion (García García, 2004).

In order to understand these socio-environmental impacts, we briefly describe some of these mining structures. Thus, the "pantanos" are the mining tailings swamps from the differential flotation washings, which were deposited on land instead of being dumped directly into the sea. They consist of very fine powdery materials (<180 microns) and are easily washed away by rainwater and can also be dispersed by the wind and inhaled by the population. In addition, these "pantanos," as they remain humid in the interior, have the risk of sliding in case of seismic movement or torrential rains. In October 1973, as a result of torrential rains, one of the tailings swamps located on the outskirts of La Unión collapsed and swept away several infrastructures, causing the death of one person. The collapse of these mining tailings swamps is relatively frequent, with serious impacts on the environment and the safety of the population. The most recent one we know of occurred in Brazil, in November 2015, when the Samarco (Vale and BHP Billiton) mining dam burst and flooded 70 km of the valley, inundating the town of Bento Rodrigues, in the city of Mariana, in the state of Minas Gerais, where 158 homes were destroyed, with some 600 people living there. Other cases in Europe are also relevant, such as the accidents in Aberfan (Wales, 1966), Stava (Italy, 1985), Baia Mare and Baia Borsa (Romania, 2000), Kolontár (Hungary, 2010). Or the one that occurred in Aznalcóllar (Spain, 1998), where the wall of the mining tailings swamp broke, and more than six million cubic metres of sludge and acidic water from the differential flotation processes of pyrite affected more than 4,500 hectares in the province of Seville, also affecting the Doñana Natural Area.

The "vacies" or "terreras" are formed by the accumulation of earth from the excavation of mining cuts. They are made up of material simply piled up by the direct dumping of earth from the extraction of the Earth in the cuts; they are easily eroded by rainfall and, in some cases, have caused landslides to adjacent areas and can be transported over greater distances through the ephemeral channels. In this group of mining remains structures, the remains of mineral concentrators by gravimetric methods, from "palanquines" or "rumbos" (local name of the Round-buddle system) are included, which are mainly made of piled earth, with few remains of chemical reagents, as they were not used in these concentration processes. In addition, in this group of mining remains structures, alterations of the original minerals that can affect the health of living beings are also common – for example, the generation of acidic waters, which originate from the spontaneous oxidation of pyrites and other sulphides associated with them. This is characteristic of metal ore mines and, in general, of any mine with sulphide-rich tailings.

The open mining pits or "cortas mineras" are open-cast workings, characterised by their truncated cone shapes. These large excavations are the main sources of the material that form the new mining landscape modelling. According to Martínez-Pagán et al. (2011), the presence of unconfined mineral debris represents one of the main risks for dust emission from abandoned mining areas, as these fine materials can be easily suspended by the wind or transported by surface and underground runoff once leached. The rate of deposition of these contaminants depends on several factors related to the origin of the trace elements, climate and wind events, as well as the characteristics of the receiving landscape, in relation to surface roughness, etc. so there is a great deal of uncertainty as to the area that could be affected. Nevertheless, there is more consensus on the effects of the presence of airborne particles from metal mines on the health of people and other living beings living in these environments. According to some studies reported in Sánchez Bisquert et al. (2017), the formation of particulate matter with a diameter between 2.5 m^{-6} and 10 m^{-6} generates important human health problems, both in the short and long terms, as they can be inhaled or ingested and absorbed by the body, depending on their bioavailability.

Impacts of mining on the Mar Menor lagoon

The pollution of the southern basin of the Mar Menor lagoon by mining activities was already highlighted in Jacqueline Simoneau's doctoral thesis in the speciality of Marine Geology and Sedimentology, carried out at the Centre de Recherches de Sédimentologie Marine de Perpignan (France) in 1970.

In addition to the presence of the various mining structures, as those aforementioned, the silting up of the southern area of the Mar Menor lagoon due to dumping over decades should be pointed. In this area, the sediments show symptoms of primary contamination, due to the fact that it is an area where mining remains were dumped, with the existence of soils with an acid pH and a high soluble metallic load. Secondary contamination processes may also be observed, with contributions of mining sediments due to the action of coastal dynamics, and tertiary contamination due to runoff and water from the ephemeral channels that flow into the lagoon (Martínez Sánchez & Pérez Sirvent, 2013). These sediments have also had an impact on infralittoral communities, as well as an increase in turbidity due to suspended solids. It should be noted that, during the deposition process, mining waste occupies the seabed, with the presence of sands with high silica and mineral contents, in addition to the sulphides that were the object of concentration in the differential flotation scrubbers and the remains of chemical products used in the differential flotation processes (cyanides, sulphuric acid, copper sulphate, xanthate, etc.).

Particularly relevant for the health of the population is the conversion of the sulphides of many of the original minerals into sulphates which, being soluble, are more easily assimilated by the organism, as Martínez Sánchez and Pérez Sirvent point out (2013, p. 318): "In the mineralised zone, the final stages of the supergene alteration process lead to a large number of sulphates with different degrees of hydration, which appear as saline efflorescences and which can retain metals released during the oxidation process". Not only is their direct presence important, but these saline efflorescences, which by capillary ascent reach the soil surface, are more easily accessible, directly by contact and/or ingestion and indirectly by being transported by air or water.

A very important effect is, as already indicated, that these sulphides in contact with air are transformed into sulphates, this being a relevant phenomenon since these sulphates are soluble and therefore bioavailable, unlike the initial sulphides which are not soluble, as explained by Martínez Sánchez and Pérez Sirvent (2013):

> The greatest risks are in surface materials, subject to supergene alteration processes, fine textured, with acid pH (…) The most important route of

> exposure is the ingestion of solid particles, given the characteristics of the material, followed by dermal and inhalation (…) Another important aspect is that of upward washing processes that must be controlled, as this is one of the important routes of soluble metals. The surface layer (less than 1 metre) contributes the most soluble metals to groundwater, so it is essential to prevent their oxidation.
>
> (p. 331)

In addition, the Mar Menor receives waste from agricultural activities (increasingly numerous and intensive from the Campo de Cartagena drainage basin) by wind transport, as well as water laden with waste from agricultural and livestock farming activities via surface and subsurface runoff. The lagoon also receives surplus water from agricultural irrigation, laden with nutrients and phytosanitary products, and the remains of desalination plants containing brine. These processes of wind transport and surface and subsurface runoff are fundamental for the arrival of the remains of mining and metallurgical activities, laden with heavy metals and metalloids and other potentially toxic elements, to the southern basin of the Mar Menor from the Sierra Minera of Cartagena-La Unión. According to Alcolea Rubio (2015):

> With regard to the transport, dispersion and deposition mechanisms of Ni, Cu, Zn, As, Cd and Pb, the mass balance of geo-available trace metals in the Mar Menor revealed that wind erosion transferred 81% of the total input of metals from the Sierra Minera, groundwater contributed 16% and surface run-off 3%. (…) All this showed that the dispersion of these pollutants is not controlled by inland waters (surface runoff and groundwater).
>
> (p. 113)

In this line, Sánchez Bisquert (2017) determined the average annual level of total deposition recorded in the study area, the Sierra Minera de Cartagena-La Unión, at 42.0 g/m^2/year, which represents a value above normal for most semi-arid areas. This means that more than 4.6 Mt/year are eroded in the more than 11 km^2 of the Sierra Minera slope to the Mar Menor lagoon loaded with mining-metallurgical waste.

In the specific case of the installations located on the north side of the Sierra Minera of La Unión and Cartagena, the waste accumulated in these deposits is transported to the Mar Menor through the existing ephemeral channels in the area, called "ramblas", such as Rambla del Miedo, Rambla de Las Matildes, Rambla del Beal, Rambla de Ponce and Rambla de La Carrasquilla (Figure 5.1). This contribution of heavy metals to the Mar Menor lagoon is a serious socio-environmental problem, in addition to the fact that these deposits can cause risks to the safety of people and property, due to the structural instability they produce together with the different elements that make them up (drainage chimneys, wells, drains, etc.).

As shown in Figure 5.1, the areas where the abandoned mining waste facilities are located are connected to the Mar Menor by these aforementioned five large ephemeral channels through which sediments and mining waste reach the lagoon.

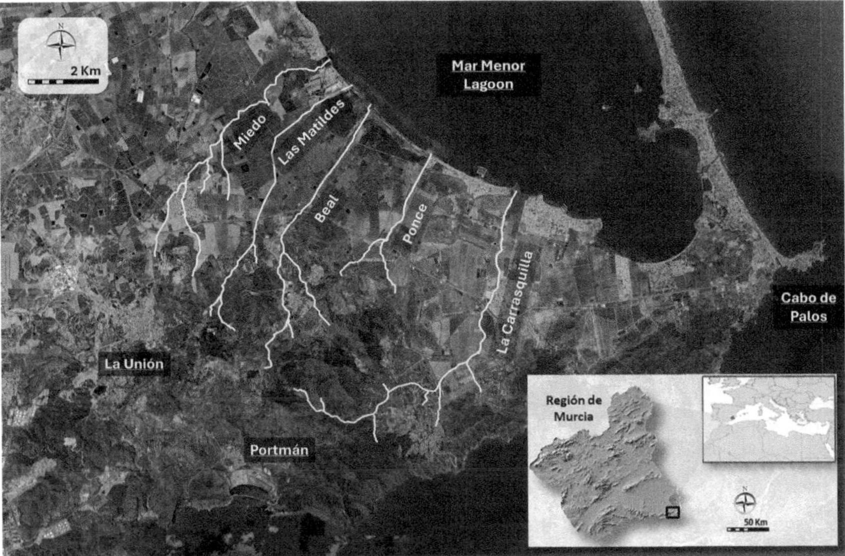

Figure 5.1 Location of the study area and main ephemeral channels.
Source: Authors' own work.

What actions should be considered to address the existing risks?

Arango Aramburu and Olaya (2012) argued that when a mining area is abandoned without remediation of environmental damage, and when this damage poses a risk to the population, there is an obligation to remediate or compensate. Thus, in view of this situation, and with the aim of reducing the socio-environmental risks to the Mar Menor, to which the metals and metalloids present in the Sierra Minera are transported via the existing ephemeral channels, actions must be considered to prevent them, both at the source of this waste and in its transport channels, thus seeking to comprehensively address the existing problem.

Based on various complaints and demands from society and environmental groups that are relevant in the Sierra Minera of La Unión-Cartagena, there are abandoned installations and contaminated soils that cause heavy metals to be dragged into the Mar Menor lagoon, the Project for the environmental remediation of mining waste and sites affected by mining in the municipalities of Cartagena and La Unión was drawn up. The Ministry for Ecological Transition and the Demographic Challenge (MITERD) presented in November 2021 the "Framework of Priority Actions to Recover the Mar Menor" (MAPMM), with the aim of addressing and intervening in the main causes that have generated and motivated the state of eutrophication and the ecosystemic crisis suffered by the Mar Menor; Specifically, *in its measure 2 is collected the aim of aim of addressing and intervening in the main causes that have generated and motivated the state of eutrophication and the ecosystemic crisis suffered by the Mar Menor.*

Table 5.1 Action groups and surface to be restored

Group no.	*Name*	*Surface to be restored (ha)*
1	Urban area of El Llano del Beal	17.57
2	Secondary school and special education school in La Unión	8.35
3	Los Benzales	8.87
4	Brunita	67.77
5	El Descargador	59.89
6	Cuesta de Las Lajas	8.53
7	Los Pajarillos-Marqués de Péjar	29.31
8	Peña del Águila	35.76
9	La Torrecica	47.25
		Total: 283.29

Source: Authors' own work, adapted from the Project for the environmental remediation of mining waste and sites affected by mining in Cartagena and La Unión.

Also in November 2021, the Council of Ministers, at the proposal of the MITERD, declared certain works for the protection and environmental recovery of the Mar Menor to be of general state interest. This declaration of urgency was justified by the critical situation of the lagoon and would allow the immediate start of the development and execution of some urgent and priority projects to halt and reverse the serious ecological deterioration of the environment.

Based on the inventory of waste facilities, implemented by RD 975/2009, in compliance with Directive 2006/21/EC on the management of waste from extractive industries, the actions for the recovery of soils and areas affected by mining in the Mar Menor basin were established. The aim of these actions is to reduce soil and groundwater pollution and prevent surface runoff of hazardous mining waste that could reach the Mar Menor. The purpose of the entire project is the drafting and technical definition of the actions for the restoration of abandoned mining waste facilities and the restoration of areas affected by mining in the area of influence of the Mar Menor in the Sierra Minera of La Unión-Cartagena basin.

These actions are aimed at the restoration of dangerous abandoned mining sites and the restoration of areas affected by mining in the area of influence of the Mar Menor. The aim is to prevent mining waste laden with heavy metals and metalloids from entering the Mar Menor through the ephemeral channels, as well as to prevent the aeolian dispersion of this waste into the lagoon and nearby urban environments, in addition to carrying out landscape and environmental regeneration of the area. The 35 mining and metallurgical waste facilities located in the area have been grouped into the following nine action groups, shown in Table 5.1.

Therefore, the objectives pursued through the implementation of the actions included in the project are based on the following:

1 To reduce the environmental risks associated with abandoned mining waste facilities, mitigating the contribution of mining waste to ephemeral channels and its transport through these to the Mar Menor.

2 To reduce wind pollution by dispersion of existing waste to nearby urban centres.
3 To increase the geotechnical and geochemical stability of the facilities.
4 To restore the relief to more natural forms that favour plant restoration and the regulation of runoff, while, at the same time, hindering erosion phenomena.
5 To implement vegetation cover that allows integration and enhances landscape restoration.

Among the possible alternatives to be used in the restoration, those that offer the best technical solution to achieve the mitigation of the polluting effects, as well as the environmental restoration of the mining waste facilities, have been considered and chosen, taking into account both their cost/benefit and the implementation time and expected results. The alternatives considered are mainly oriented towards the analysis of the following aspects:

- The topographical remodelling of the land.
- The sealing system for mining waste facilities and affected sites.
- The restoration of the vegetation and its integration into the landscape.

In order to carry out the environmental remediation works, the topographical remodelling of the land will have to be carried out, with the regularisation of platforms and the laying of existing slopes at the different mining waste facilities and the affected sites. In some cases, it is planned to remove existing material from other areas of the same site to reuse it as fill material. On other occasions, the remodelling will be minimal, the aim being to create a surface suitable for the treatment selected for each case.

The alternatives selected for the main treatment of the soils contaminated by mining waste located in the different installations and sites within the area of action are listed below, from the highest to the lowest degree of intervention:

1 *Encapsulation "in situ"*. It consists of adapting the existing tailings deposits and covering them with a series of layers to guarantee the waterproofing and stability of all the tailings contained. These layers will be made up of materials such as clay, gravel and topsoil, in addition to the installation of geomembranes with high density polyethylene (HDPE) sheets and geotextile to protect and separate the different layers. The waterproofing function of the set of confined waste after the construction of these layers is complemented by the creation of breakwaters to protect the slopes in order to ensure the structural stability of each deposit, as well as the creation of a drainage network to channel rainwater runoff, control erosion and increase the useful life of the project.

   Finally, the surfaces will be revegetated to improve environmental remediation and to favour their integration into the landscape.
2 *Technosoils*. These are artificial soils created by mixing different raw materials in different proportions adapted to the substrate where they are applied, with the aim of interacting with the waste present, favouring their physical-chemical

stabilisation and the creation of a new substrate capable of preventing the mobilisation of metals and metalloids. The materials with which these technosoils are made can be manure, pruning and biomass waste, compost, biomass combustion ashes, clays, dust and sawdust from calcareous rocks, marble, porphyry, etc. A good alternative is sought for the recovery of soils degraded by mining-metallurgical activities, based on the use of these technosoils to improve the physical-chemical properties and conditions of the soil, controlling the processes of environmental degradation and trying to reconstruct the landscapes affected by mining activities.

3 *Phytoestabilisation.* It is based on the interactions between plants, soil and microorganisms in those areas where the concentration of pollutants is lower. And also to complete the environmental remediation of the whole and landscape integration. Different plantations would be used to create a stable vegetation cover, to minimise the mobility and bioavailability of existing pollutants in the soil, reduce erosion and surface runoff, as well as trying to avoid the production of secondary pollutants.
4 *Bio-engineering.* It favours stability and avoids erosive processes on surfaces and slopes in the planned action areas. Different techniques will be used, such as "fajinas, barricades" or the planting of live stakes in breakwaters and slopes, which will be used for the restoration of vegetation and the landscape integration of areas degraded by mining and metallurgical activities.

These works, included in the Actions for the restoration of dangerous abandoned mining sites and restoration of areas affected by mining in the area of influence of the Mar Menor, included in the MAPMM, to be carried out by the MITERD, as well as other actions in ephemeral channels in the Sierra Minera, are part of a process of restoration and environmental adaptation of the Sierra Minera, by the Confederación Hidrográfica del Segura, an autonomous organism of the State General Administration also dependent on the MITERD, are part of a process of restoration and environmental adaptation of the Sierra Minera of Cartagena-La Unión. We also consider that these works should also continue on the Mediterranean Sea side, as the socio-environmental problems and the effects on living beings are similar throughout the Sierra Minera. Therefore, it is important to restore the Sierra Minera as a whole. It is also urgent because the effects on the health of people and living beings caused by the waste from mining and metallurgical activities are continuous, as they are real physical and chemical reactors.

Final remarks

The sum of the impacts of mining, agriculture and urban development in the Mar Menor area during the last decades has clearly affected the entire socio-ecological systems (Conesa & Jiménez-Cárceles, 2007; Guaita García, 2021).

Although many studies and actions are being carried out in the system, the complexity of its overall functioning and the slow, and even scarce, compliance in the implementation of the recovery actions of the areas affected by the mining -and other activities, should be highlighted.

These actions, in addition to generating employment and improving the health and living conditions of the area's residents, would favour the development of other economic activities that would make the area more dynamic. One of these could be the implementation of a new, more sustainable tourism development model based on ethnography and nature, serving as a complementary offer to the tourist areas of La Manga and the Mar Menor where the development model has been almost exclusively sun and beach tourism. In this new design, environmental conservation and restoration, together with the rich cultural heritage of the surroundings, would be fundamental axes for sustainable development based on the values of the natural and geological-mining heritage of the area, different from the more conventional models of tourism developed in the surroundings, especially in the area around the Mar Menor.

References

Alcolea Rubio, L. A. (2015). *Geoavailability of Ni, Cu, Zn, As, Cd, and Pb in the Sierra de Cartagena-La Unión (SE Sapin)*. Doctoral dissertation. UPCT.

Arango Aramburo, M., & Olaya, Y. (2012). Problemática de los Pasivos Ambientales Mineros en Colombia. *Gestión y Ambiente, 15*(3), 125–133.

Banos-González, I., & Baños Páez, P. (Eds.) (2013). *Portmán; de El Portus Magnus del Mediterráneo Occidental a la Bahía Aterrada.* Editum.

Banos-González, I., Baños Páez, P., Pérez Cutillas, P., & Esteve Selma, M. A. (2017). Análisis de las propuestas de los actores sociales en la recuperación ambiental de la Bahía de Portmán (Región de Murcia). Nuevas perspectivas para un desarrollo sostenible. *Cuadernos de Turismo, 40,* 135–154.

Baños-Páez, P. (2023). Le regard du sociologue. Usages et abus dans la utilisation de la mer. Les cas de la baie de Portmán et la Mar Menor, dans le sud-est de la péninsule ibérique, en Méditerranée occidentale. In Bereni, A., Ricard, P., & Seddik, W. (Eds.), *Regards croisés sur la nécessaire conciliation entre activités humaines dans les eaux européennes. Conflits d'usage en Mer* (pp. 83–108). Editions A. Pedone.

Belmonte Serrato, F., Romero Díaz, A., & Moreno Brotóns, J. (2010). Contaminación ambiental por estériles mineros en un espacio turístico en desarrollo, la Sierra Minera de Cartagena-La Unión (sureste de España). *Cuadernos de Turismo, 25,* 11–24.

Conesa, H. M., Faz, A., & Arnaldos, R. (2006). Heavy metal accumulation and tolerance in plants from mine tailings of the semiarid Cartagena-La Unión mining district (SE Spain). *Science of the Total Environment, 366,* 1–11.

Conesa, H. M., & Jiménez-Cárceles, F. J. (2007). The Mar Menor lagoon (SE Spain): A singular natural ecosystem threatened by human activities. *Marine pollution bulletin, 54*(7), 839–849.

García García, C. (2004). *Impacto y riesgo ambiental de los residuos minero-metalúrgicos de la Sierra Minera de Cartagena-La Unión (Murcia-España).* Doctoral dissertation. UPCT.

Guaita García, N. (2021). *Evaluación integrada de sostenibilidad en un sistema socio-ecológico complejo del litoral mediterráneo mediante un proceso de investigación inter y transdisciplinar.* Doctoral dissertation. University of Alcalá de Henares.

Levasseur, P., Erdlenbruch, K., Gramaglia, C., Bento, S., Fernandes, L. & Baños Páez, P. (2022). Does pollution perception lead to risk avoidance behaviour? A mixed methods analysis. *Review of Social Economy,* 1–27. https://doi.org/10.1080/00346764.2022.2030542

Martínez-Pagán, P., Faz, A., Acosta, J. A., Carmona, D. M., & Martínez-Martínez, S. (2011). A multidisciplinary study for mining landscape reclamation: A study case on two tailing ponds in the Region of Murcia (SE Spain). *Physics and Chemistry of the Earth, 36*(16), 1331–1344.

Martínez Sánchez, M. J., & Pérez Sirvent, C. (2013). Diagnóstico y Recuperación de la Contaminación del Suelo en Portmán y Sierra Minera. In Banos-González, I., & Baños Páez, P. (Eds.), *Portmán; de El Portus Magnus del Mediterráneo Occidental a la Bahía Aterrada* (pp. 313–343). Editum.

Romero Díaz, A., Belmonte Serrato, F., & García Fernández, G. (2011). Erosión y efectos de la minería en la Sierra de Cartagena-La Unión. In Hernández Bastida, J. (Coord.), *Recorridos por el campo de Cartagena. Control de la degradación y uso sostenible del suelo.* (pp. 147–178). Instituto Euromediterráneo del Agua (IEA).

Sánchez Bisquert, D. (2017). *Caracterización y dispersión del material aerotransportado en el entorno de la antigua zona de minería metálica de Cartagena-La Unión: Impactos y riesgos*. Doctoral dissertation, UPCT.

6 Cultural landscapes' deterioration as losing one's aesthetic environment

The case of Mar Menor

María José Alcaraz León[1]

Introduction

The category of "cultural landscape" helped to bring to the foreground the importance of perceiving human-modified environments as an expression of the cultural values that emerge from the interaction between cultural communities and their natural environments. As "landscapes", cultural landscapes prominently have aesthetic character. This is twofold: first, the processes involved in the formation of cultural landscapes may constitute new salient aesthetic properties. Second, some of those aesthetic properties may play a significant role in the constitution of the cultural landscapes' identity, becoming hallmarks of its singularity and overall value. Although this aesthetic dimension of cultural landscapes is directly connected to the physical transformation of the environment in its relation to a cultural form of life, in some cases, the cultural print is not physically manifest. In these cases, the aesthetic character of the cultural landscapes is uniquely tied to the cultural immaterial relations – such as historical events or religious beliefs – that the cultural group has with their particular environment. These relations crystallize into a perceptual experience of the environment that is aesthetically rich. This is the case, for example, of Tongariro's National Park or Fuji Mount, where the landscape is perceived as charged with divine connotations. In these cases, the aesthetic appreciation of cultural landscapes involves taking into consideration a particular's group relation to it and the environment's cultural meaning for this particular group. As a result, the aesthetic character of these cultural landscapes is strongly community-bound: its perception is inseparable from the activities, stories, attachments, etc., that the inhabitants have sustained in relation to that environment. On the other hand, this community-bound character may seem, in principle, to conflict with some default assumptions about aesthetic value: specifically with an understanding of Aesthetics as an activity where the notion of disinterest and the aspiration to a universal or at least intersubjective judgement agreement are central (Kant, 2000).

In this chapter, I would like to explore the apparent tension between the alleged universal character of aesthetic value and the community-bound dimension of the aesthetic appreciation of cultural landscapes by focusing on the case of Mar Menor. Although this environment has not been officially categorized as a cultural landscape, it possesses many of the features that make for that status.[2] As a result,

DOI: 10.4324/9781003489078-7

paying attention to its aesthetic character under the light of this category may help us to better grasp its aesthetic complexity and to acknowledge the dependence of this value in the fragile equilibrium that the different natural and human components have generated.

In this sense, I would like to propose that the Mar Menor case helps us to envisage the importance of the varied relationships that communities living around a natural environment have historically entertained and that have been crucial to this environment's aesthetic values. Furthermore, I would like to motivate the idea that this circumstance, far from compromising the aspiration to universality proper to the aesthetic judgment, should be taken as the condition for its crystallization. Finally, given that Mar Menor is experiencing a significant transformation due to various economic activities, I will show that it is a vivid example of the experience of losing one's place or aesthetic environment and I will explore some of the consequences this experience may have.

Cultural landscape and aesthetic value

The term "Cultural landscapes" was coined in 1992 to identify and protect some environments whose values result from a particular form of human interaction with those environments. As it is characterized in the UNESCO Operational Guidelines,

> Cultural landscapes are cultural properties and represent the "combined works of nature and of man" designated in Article 1 of the Convention. They are illustrative of the evolution of human society and settlement over time, under the influence of the physical constraints and/or opportunities presented by their natural environment and of successive social, economic and cultural forces, both external and internal.[3]

Cultural landscapes are thus unique manifestations of certain forms of nature/culture interactions. Their uniqueness very much depends on the particular development of a form of life whose very existence and practices cannot be disentangled from a particular environment. This relational nature of cultural landscapes generates a fragile equilibrium between the natural conditions that make certain human practices possible in a particular environment and the contribution that those practices have in preserving certain natural dynamics.

Resulting from these interactions, as well as from the peculiar natural conditions that facilitate them, cultural landscapes are characterized by a great variety of values, including aesthetic value. In my approach to these environments, I will focus mostly on the aesthetic value of cultural landscapes with the further hope of clarifying how this value becomes entrenched with other distinctive aspects and activities of these environments. I hope to clarify the complex nature of this value – insofar as it emerges from the dynamic interactions characteristic of cultural landscapes – and to show how it partly constitutes cultural landscapes' overall character.

We can begin by noticing that the aesthetic dimension of cultural landscapes is one of its main defining features. First, the very expression of cultural landscapes refers to the environment as a "landscape"; that is, as a delimited area or environment which is perceived as an aesthetic whole or unity (Berque, 1998; Roger, 1978). Thus, at the very origin of the concept of landscape, we find the idea of an aesthetically imbued perceptual experience and, hence, cultural landscapes inherit this aesthetic dimension.[4]

Second, this aesthetic dimension is also acknowledged by UNESCO as one of the reasons that make for an environment a cultural landscape. Among the ten criteria stipulated for the recognition of a site as a heritage candidate (see https://whc.unesco.org/en/criteria/), criterion (vii) "to contain superlative natural phenomena or areas of exceptional natural beauty and aesthetic importance" directly refers to a site's aesthetic value as one of the reasons for protecting it as human heritage. Other criteria (such as (i), (ii), (iv), or (vi)) are also directly or indirectly related to the artistic and aesthetic character of an environment as a reason for its categorization as Cultural Landscape. Thus, the aesthetic value of cultural landscapes is generally acknowledged, at least among the referred criteria, as one of their constitutive values.

Despite this, the study and understanding of the aesthetic value of cultural landscapes are still underdeveloped. Although some research has been done on the particular aesthetic character of certain cultural landscapes (Kibebew, 2021; Pechanec et al., 2020), there is less work addressing a general characterization of this value in relation to the particular nature of cultural landscapes and their complexity. In this sense, the distinct aesthetic character of cultural landscapes has received comparatively less attention than other environments' aesthetic character, such as natural environments or urban environments. Even within the domain of Environmental Aesthetics, the literature on the aesthetic appreciation of natural environments has dominated the main debates.[5] Only recently more attention has been paid to mixed or human environments among which cultural landscapes could be included.[6]

Still, we can at least point to three distinctive aspects of the aesthetic value of cultural landscapes as instances of mixed environments.

First, its origin is inextricably linked to the sustained cultural activities performed in that environment and so, it is a kind of aesthetic print or crystallization of a culture's activities and relations to that environment. A good example of how human sustained activity has moulded the environment producing landscapes of remarkable beauty is the rice terraces of Philipines' cordilleras in the Ifugao Province (https://whc.unesco.org/en/list/722). This landscape was one of the first categorized type 2a cultural landscapes. Tellingly characterized as "stairs to heaven", the stratified hillsides of the mountains caused by the long-standing rice farming constitute a harmonious pattern of "unparallel beauty" (sic). Its characteristic aesthetic appearance and value result from the persistent interaction between the Ifugao people and the environment. Thus, to fully grasp those forms as what they are, it is necessary to consider how both natural conditions and human activities have contributed to their formation.

In this respect, the aesthetic value of cultural landscapes is different from the aesthetic value of natural environments, on the one hand, and urban environments, on the other. In natural landscapes, the aesthetic value mostly derives from the natural elements, forces and dynamics that contribute to a particular environment's overall formal pattern; in urban landscapes, the aesthetic value is mostly derived from and is an expression of, the more or less intentional and dynamic human activities that have constructed the environment. Understanding the relations that are at the origin of the aesthetic value of cultural landscapes requires paying equal attention to the human activity (broadly construed) and to the natural conditions of that environment. Moreover, it requires paying attention to the particular relations between human activities and the natural environment. In this respect, a remarkable aspect of the characteristic relations of some organically cultural landscapes is that they often produce changes in the natural dynamics that can in turn generate further ecological diversity and beneficial symbiotic relations. As I will try to illustrate with the Mar Menor case, some of the economic activities that alter a site's natural biodiversity, such as the exploitation of coastal salt flats, can generate new ecological environments that are apt for plant and animal species whose presence enriches, in turn, its ecological and aesthetic value.[7]

Second, the interactions that are at the basis of the aesthetic character of a particular cultural landscape do not always leave a physical print that can be easily recognized through bare perception. This is especially apparent in the case of the third subcategory of cultural landscapes: associative cultural landscape. For example, the perception of an environment as a site of religious significance or as an important historical fact or event can imbue the environment with some expressive properties or permeate the perceptual experience of the environment with historical or religious meaning, which, in turn, endows it with an aesthetic character. Consequently, grasping this aesthetic character requires being aware of the historical or religious beliefs that infuse that perception or adopting a particular perspective that makes that kind of perception available. Considering a particular example can be illuminating. The Tongariro National Park was the first inscribed associative cultural landscape. This site had been previously inscribed as a natural landscape for its geological and ecological values, but due to its religious significance to the Maori people, it was re-inscribed as a cultural landscape in 1993. The characteristic volcanic environment of this park was, for the Maori people, imbued by religious connotations and mythological beliefs about their origin as a community. In this sense, their cultural identity is very much associated with the environment and, hence, this aspect partly constitutes it as an experienced environment. This character is further manifested in distinctive aesthetic perceptions of the environment, such as the overwhelming or solemn character of the volcanos once conceived as gods' household.

Third, there is an aesthetic dimension in the very activities and practices performed in relation to the environment that may be appreciated on their own. For example, as I will try to illustrate with some of the practices that are characteristic of the Mar Menor, there is an aesthetic dimension related to the practices of salt extraction or certain traditional fishing methods that can reflect into the environment's aesthetic perception. These activities have their aesthetic character which is, as the aesthetic value of the cultural landscape in which they take place, intimately linked to the characteristics of the environment while preserving their distinct character.[8] For example, certain activities possess a certain rhythm or promote certain forms of attention that are constitutive of their aesthetic dimension. For example, using a particular set of fishing tools generates patterns of interaction with the environment. They constitute certain forms of moving through the space and attending to it that are partly generated by the tools themselves, but also by the environment within which they operate. As a result, the aesthetic character of these activities manifests the co-dependence characteristic of manufactural forms of engaging with the environment.[9]

Consequently, characteristic patterns of interaction that constitute the environment as a case of cultural landscape generate forms of attention and perception within which their singular aesthetic dimension materialises. In this sense, the aesthetic character of cultural landscapes also contains this dimension of aesthetic attention that is facilitated by the very activities characteristic of a culture.

Now, the aforementioned complexity and characteristic origin of cultural landscapes' aesthetic value can also be considered as something that problematizes the universal availability and recognisability of that aesthetic value. Given its intimate connection to certain cultural practices and local forms of interaction with the environment, there could be a certain tension between the local conditions required for the recognition of that aesthetic value and the traditional understanding of aesthetic value as a form of value whose availability only requires certain shared human capacities and a disinterested form of attention to the formal aspects of the object contemplated. In 'Section Local dependency and universal accessibility of the aesthetic value of cultural landscapes', I will try to show that the distinctive character of the aesthetic value of cultural landscapes does not conflict with these aspirations. Paying attention to how aesthetic value emerges in the characteristic context of cultural landscapes, and within the relations that typically constitute it, can throw some light onto some neglected aspects of aesthetic valuing in general.

Mar Menor as an example of the aesthetic richness of cultural landscapes

Mar Menor is a singular coastal lagoon located on the South East coast of Spain (Murcia). It is connected to the Mediterranean by three natural channels or "golas" facilitating the lagoon's water renewal.[10] Its waters are saltier, warmer and more transparent than the Mediterranean's. This has facilitated the presence of a distinctive natural environment which has, in turn, enabled the development of certain human activities – such as fishing and salt production – which, at least until the 70s of the past century, coexisted in a singular equilibrium with this environment.

Current scientific research on the ecological state of Mar Menor acknowledges the impact of both traditional and more recent human activities on the current ecological state and dynamic of this unique coastal lagoon (Ruiz et al., 2020). Some of these traditional activities, like "encañizadas" (a traditional fishing method) or the salt extraction in San Pedro del Pinatar, are mentioned in some of the last experts' reports as activities that have traditionally had beneficial effects on the ecological sustainability of the lagoon and whose subsistence would be beneficial for the lagoon's ecological equilibrium. Other, more recent activities, like the increase of touristic resorts and the intensification of agricultural activity in Campo de Cartagena (with the corresponding increase of chemical inputs into the lagoon), have proved to alter considerably the biological equilibrium of the lagoon[11] and to contribute to the deterioration of significant portions of its flora and fauna. These activities have also had a great impact on the lagoon's singular hydro dynamism in certain parts of its coast. For example, the widening during the 1970s of one of the natural channels that connected Mar Menor and the Mediterranean, the Estacio, the construction of breakwaters in some key points of the coast and the creation of new artificial beaches have altered the former hydrodynamics of some key coastal areas (especially in the western part of the lagoon). These activities have transformed the lagoon's basins and have altered the former habitats of its flora and fauna.

Thus, while human activity has been for centuries one of the conditions of the biodynamics of the Mar Menor, it has not had the damaging effect that has been observed in the last 50 years. Hence, the Mar Menor has been for centuries an environment that has hosted many human practices that far from deteriorating it have contributed to its biological equilibrium. In this sense, like other characteristic type 2 cultural landscapes, it has been partly moulded by those activities. In this section, I would like to comment on two of these traditional activities to show how their presence has not only contributed to the lagoon's ecological stability but also to its cultural significance and aesthetic value.

These activities are the coastal salt flats of San Pedro del Pinatar and Marchamalo and the traditional fishing method of "encañizadas". As mentioned, the benign environmental effects that the presence of these activities has had in the Mar Menor ecological equilibrium are well-documented.[12] However, the study of the cultural significance of the traditional activities and their contribution to the aesthetic and cultural values of the environment of Mar Menor is less developed. In this chapter, I would like to emphasize how these activities embody the characteristic pattern of interaction of cultural landscapes and have contributed to the emergence of singular environmental and cultural values.

Of the two saline locations in Mar Menor, San Pedro del Pinatar and Marchamalo, only the former is currently in use and is under the protection of several national and regional plans to preserve its biodiversity and cultural elements.[13] The practice of salt extraction in the Mar Menor salines dates back to the Roman occupation of the peninsula and continues until today in the area of San Pedro. As it is characteristic of cultural landscapes type 2, the resulting landscape cannot be fully understood without taking into consideration both the natural conditions that have made it possible and the impact of human activity on that environment.[14]

Thus, the salt extraction activity is mostly due to the hypersaline character of the Mar Menor, together with a very warm climate, which enabled the evaporation of the water in the different ponds, and the regular presence of wind currents that facilitated the use of windmills to transport the water into the ponds. On the other hand, the current characteristic biodiversity of the area is directly linked to the practice of salt extraction and to the environmental conditions that it generates.

In this sense, its rich biodiversity and interest could not be fully explained without acknowledging the importance of this practice. The salt ponds and channels provide the conditions for several flora and fauna species which are specially adapted to this kind of environment. Simultaneously, the different installations and processes involved in the salt extraction – such as former windmills, or the colour patterns generated by the salt ponds and the salt mountains – generate a landscape that adds its aesthetic character to this ecological diversity.

Within this area, we also find another activity that possesses great cultural (López-Martínez & Espeso-Molinero, 2020) and ecological values (Ballesteros Pelegrín et al., 2018). This is the fishing method referred to as "Las Encañizadas" (fishing in the reeds). It has been typically used in the shallow waters of the channels or "golas" that connect Mar Menor and the Mediterranean. This is not only one of the points through which the two seas are communicated, but it is also where there is a seasonal transit of different fish species. Many fish either enter the Mar Menor to spawn or leave for the Mediterranean once they have reproduced or reached a certain size. This phenomenon together with the shallow waters facilitates the development of the special fishing technique that has been practised in the area for centuries (at least since the Arabic period of the Spanish Peninsula). This type of fishing is carried out by sticking rods and nets into sandy bottoms forming a funnel network where the fish, mostly mullet and bream, are trapped. A remarkable aspect of this fishing method is its sustainable functioning (Ballesteros Pelegrín et al., 2018). The network is made of "encañizadas", reed walls that allow for the smaller specimen to return to the sea and develop its full biological cycle. As a result, the cultural and ecological values of this traditional fishing method go hand in hand. On the one hand, the practice is possible due to the existence of these shallow channels connecting the two seas and to the seasonal fish transit and, on the other, it contributes to the maintenance of the fish population by only capturing the fish that will be apt for human consumption, freeing the smaller specimen. Further, the materials used in the funnel network are the reeds obtained in that very environment and that are regularly cut to avoid the obstruction of the water entry from the dry river beds. Thus, it is an example of a sustainable use of local resources that avoids their overexploitation.

In this sense, both the practices related to the salt ponds and the "encañizadas" are clear examples of the kind of human activities that exemplify the cultural landscapes' characteristic forms of interaction between the natural environment and human activity. Both are made possible by the particular environmental conditions that are found in that area, such as the high salt density in the Mar Menor, in the case of the salines and the natural shallow water channels and seasonal fish migrations, in the case of the "encañizadas"; and both are practised in a way that contributes to the area's ecological sustainability and even to the enrichment of its biodiversity.

But it is not only the felicitous interaction between human activity and the natural environment that makes for the consideration of this environment as an example of a cultural landscape. This felicitous interaction has further produced a landscape of remarkable cultural and aesthetic value, which, in turn, contributes to the area's overall value as a cultural landscape. Thus, for example, the colour pattern of the salt ponds (specially the pink and white characteristics of the salt ponds and salt heaps) together with the presence of architectural elements, such as the traditional windmills (formerly used to elevate the water from the sea into the drying ponds) and the presence of the special vegetation and fauna, composes a landscape of great beauty. In grasping these qualities, the viewer becomes aware in an aesthetically pleasing and vivid manner of the different components that make for this singular landscape and of their dynamic co-dependence.

Besides, as mentioned above, as an example of a cultural landscape, the aesthetic dimension of this environment also engages the viewer in other ways. The two referred practices generate patterns of temporal attention and spatial organization of the environment's perceptual features. Thus, for example, the different stages of salt production and extraction generate patterns of attention that make certain aspects of this singular landscape salient. These patterns of attention are reflected in the developed special vocabulary that is created to refer to properties that would have probably remained unnoticed were not for the presence of these activities.[15] For example, one of the stages of salt production is described as "salt flowering" referring to the first appearance of salt crystallizations that can be later collected. In a sense, the fact that this expression has been coined to refer to those fragments of salt crystallization invokes a form of perceiving them that draws an analogy with the characteristic forms of flowers and their beauty. Other terms, like "harvesting" the salt, remind us of one of the characteristic activities of agriculture and assimilate salt to a living product which grows from Earth.[16]

The process of cultivating the salt is also, like other agricultural activities, very much dependent on the seasons. Thus, the salt is mostly produced during the summer – when there are higher temperatures and thus the evaporation process is faster – and collected in the Fall. Thus, the appreciation of this landscape changes through the seasons, offering different atmospheres.

Similarly, the fishing practices of the "encañizadas" involve their own temporality and corresponding forms of attending to the sea's fauna activity, specially to the patterns of migration and reproduction of the different species. These patterns of attention and activity do not only favour forms of perceiving the environment which reveal their aesthetic potential, they also possess a rhythm and an aesthetic dimension of its own. For example, the salt collection, especially when it was done manually, using spades and esparto grass baskets, involved a rhythmic movement similar to other agriculture activities.

Thus, these two activities and the ways they contribute to the constitution of this singular landscape are clear phenomena that make for the cultural value of Mar Menor and, hence, for its consideration as a cultural landscape. They not only frame the perceivers' attention to the environment making them capable of accessing some properties that become especially vivid within those practices, but bear their own aesthetic character. For example, they are informed by a certain delicacy and rhythmic qualities and by patterns of attention and appreciation that are directly connected to those activities.

Local dependency and universal accessibility of the aesthetic value of cultural landscapes

Since I have characterized the overall aesthetic value of these environments as being partly constituted by the aesthetic dimension of the local practices referred to above, there may be a worry that this dimension is too tied to the locals' experience for it to be universally available. In this section, I would like to offer a reply to this worry and to show that the aesthetic value that derives from these practices is not necessarily opaque or less accessible than other aesthetic values that may be regarded as deriving from the environment's natural biodiversity. In principle, one could think that those aesthetic values that are grounded in the environment's physical and natural features can be easily acknowledged or perceived by, at least, those appreciators who are to a certain degree familiar with forms of appreciating nature. This kind of appreciator is not hard to find nowadays, especially since the ecological movements initiated in the 60s and the increased awareness of the beauty of the environment as one of its worth preserving values.[17]

This sensibility towards natural beauty is to a great extent moulded by different forms (scientific, artistic or practically oriented) of representing/describing nature. Our current appreciation is in this sense, highly influenced by the patterns of framing and focus that these representations embody and that train our later perception of the environment. I think that a similar acquaintance with the patterns of attention and engagement involved in the practices mentioned above and with the corresponding dimension associated with these practices is the route to being capable of appreciating the environment in a way that converges with the locals' characteristic aesthetic engagement. These processes do not require, I think, that the appreciators – be it the locals or the visitors – are aware of all the elements that enter into their experience or that they can tell apart clearly which aspects are, from their overall experience with an environment, properly aesthetic. As noticed, cultural landscapes possess a complex variety of values and their appreciation often reflects this complexity.

The sceptic could consider this explanation as insufficient to grant the accessibility or generality of this value. After all, the possible aesthetic value of an environment which is due to the development of certain cultural practices is contingent on the existence and development of that culture, while the aesthetic character that can be derived from the environment's natural morphology and biodiversity seems more stable or intimately linked to that environment. My response to this is precisely to question the assumption behind this idea. In my view, the sceptic wrongly assumes that aesthetic properties and value are only dependent on the physical perceivable properties of an object or environment. However, this is a contested assumption (Danto, 1996; Levinson, 2006).

The forms of interacting with an environment, be it scientific, practical or merely contemplative, always involve some form or another of attending to, and therefore organizing the environment we relate to in a certain way. This does not mean that the resulting aesthetic properties or values that become available to us are mere projections with no relation whatsoever to the environment. They are grounded in the environment's qualities, but their grasp requires a particular perspective or way of engaging with that environment.[18] In this sense, neither the aesthetic value that we can find in their more natural elements (Budd, 2002) nor the ones that derive from the activities performed in a particular environment are given to us independently of some form or another of engagement (Brady, 2023). Each becomes available within the practice that provides an organizing frame or perspective from which certain aspects become salient.

If these considerations are convincing, then there is nothing that precludes acknowledging the aesthetic value of those practices and of what these activities reveal about the natural environment. Grasping those values requires certain knowledge of or familiarity with the activities that ground them, but this is a condition that all aesthetic values share, insofar as its availability is connected to certain forms of interacting with our environment.[19]

Losing one's place

Finally, I would like to pay attention to some of the experiences that have partly motivated the social movements to protect Mar Menor from its ongoing deterioration and that can be characterized in terms of "solastalgia" (Albrecht, 2005, 2020; Albrecht et al., 2007) or the experience of losing one's place (Nomikos, 2018).

These expressions have been recently coined to refer to a set of cognitive-affective experiences that people familiar with a particular environment have in the face of its drastic transformation. Thus, solastalgia ("the distress that is produced by environmental change impacting on people while they are directly connected to their home environment") has become a common experience due to the fast transformation of human environments caused, among other human and natural factors, by the homogenization of urban spaces, Global Climate Change (GCC) effects on natural and mixed environments, or aggressive economical activities that inflict drastic transformation of the environment. The pervasive character of this experience reflects the existential significance that the places we are familiar with have for us and the overall cultural degradation that losing one's place involves (Ramsay & Han, 2012).

In the field of Aesthetics, Arto Haapala and Hans Maes (Haapala, 1999, 2005; Maes, 2022) have also paid attention to the aesthetic dimension of the existential relationship to one's everyday environment. As aesthetic agents, we are responsive to our everyday environment and we existentially constitute ourselves in part by those aesthetic relations. When these relations deteriorate or become unavailable due to a drastic alteration of the environment, there is an impact on a significant dimension of our everyday existence. The experience of losing one's place can, thus, manifest aesthetically as a form of "aesthetic melancholy" (Maes, 2023) which registers this sense of loss.

These experiences are becoming increasingly frequent among the Mar Menor local population. The fast deterioration of the environment and the awareness that certain damages cannot be completely reversed have, especially since the first big ecological crisis of the Mar Menor in 2016,[20] contributed to these experiences which register the existential significance of our natural environments and their importance for our sense of being in the everyday world.

Conclusions

The notion of cultural landscape has contributed to better understanding and protecting the complexity and values of certain environments that are constituted by a special interaction between natural conditions and human practices of special cultural value. Although the Mar Menor has not been classified as a cultural landscape, it possesses many features that could make it an example of type 2a of this category. In this chapter, I have tried to approach the aesthetic richness of the Mar Menor environment from the perspective of this category and to show how its natural conditions and the practices that have been traditionally informing the landscape are essential for the aesthetic richness that has characterized this singular lagoon. As I have tried to show, the aesthetic complexity of this environment understood as a cultural landscape resides in the characteristic forms generated by certain practices as well as by the forms of perception and attention that those very practices promote and foster. Finally, the fast deterioration of the environment due to the expansion of touristic resorts and agricultural practices and the resulting abandonment of some of the traditional activities that were characteristic of this landscape have generated a corresponding experience of loss and anxiety that some recent authors have captured in terms of new emotional responses directly reflecting the negative feelings associated with losing one's space.

Notes

1 This research has been possible as a result of the funding received by the research project "The Rationality of Taste. Aesthetic appreciation and deliberation" (PID2023-149237NB-100) (Ministerio de Ciencia, Innovación y Universidades. Agencia Estatal de Investigación) and by the Plan de recuperación, transformación y resiliencia – financiado por la Unión Europea – NextGenerationEU.

2 This lack of official acknowledgement by UNESCO is not very problematic if we take into consideration two aspects. First the notion of "cultural landscape" is relatively recent (1992) and, as a result, second, many possible environments that could be candidates for this recognition are still under consideration or evaluation. For the present contribution, it will suffice to identify some of the features that would make this unique environment a sure candidate for the cultural landscape acknowledgement.

3 Cultural landscapes encompass three further subcategories: (1) "the clearly defined landscape designed and created intentionally by man" (typically garden and parkland landscapes); (2) "the organically evolved landscape. This results from an initial social, economic, administrative, and/or religious imperative and has developed its present form by association with and in response to its natural environment." This second type further subdivides into two sub-categories: (2.a) "Relict (or fossil) landscape is one in which an evolutionary process came to an end at some time in the past, either abruptly or over a period" and (2.b) "Continuing landscape" one "which retains an active social role in contemporary society closely associated with the traditional way of life, and in which the evolutionary process is still in progress." The final category is the "associative cultural landscape". "The inclusion of such landscapes on the World Heritage List is justifiable by virtue of the powerful religious, artistic or cultural associations of the

natural element rather than material cultural evidence, which may be insignificant or even absent" (WHC, n.d., pp. 22–23) (see Art, 47, 47bis and 47ter of the https://whc.unesco.org/en/culturallandscape/#2).

4 In this sense, landscapes in general also possess a cultural dimension or are partly constituted by a certain cultural relation to the environment. As Schama pointed out in his classical work *Landscape and Memory* "Landscapes are culture before they are nature; constructs of the imagination projected onto wood and water and rock...once a certain idea of landscape, a myth, a vision, establishes itself in an actual place, it has a peculiar way of muddling categories, of making metaphors more real than their referents; of becoming, in fact, part of the scenery" (Schama, 1996, p. 61).

5 For example, in one of the first compilations about the aesthetic appreciation of human environments (Berleant & Carlson, 2007), most of the contributions address the aesthetic experience of the urban environment.

6 The main contributions in this regard (Arntzen & Brady, 2008; Berleant, 2010; Brady, 2006; Brady, 2008; Taylor et al., 2015).

7 Although the ecological alterations produced by the development of these activities are often considered positive, due to the environmental conditions they generate and the biodiversity enrichment following these environmental conditions, some environmentalists defend that any human alteration always involves some diminishment of environmental value. This view is also echoed within Environmental Aesthetics. Thus, the defenders of what is known as "Positive Aesthetics" consider that any human alteration of a natural environment always detracts from its natural beauty.

8 A good analogy of this aesthetic character can be found in the aesthetic dimension of the activities performed in a garden, where the activities possess a rhythm that generates patterns of attention that are partly determined by the natural conditions of the environment. See (Cooper, 2008). Similarly, Saito (2010, 2017, 2022) has elaborated on the idea that there is an aesthetic dimension in the very everyday practices through which we interact with our familiar environments.

9 This co-dependence may be maybe missing in the characteristic forms of technical interaction that are common after industrialization, for they are to a great extent more independent from the constraints characteristic of an environment.

10 In contrast to the two other lagoons that are often studied in comparison with the Mar Menor, the Venetian lagoon and Curonian lagoon (in the Baltic Sea), the former transparency of the waters of Mar Menor was due to the slow pace of its water renewal, which expanded for more than a year before the Estacio channel was artificially enlarged in 1974–1975 to allow maritime transit. Currently, the Mar Menor's waters renew in less than nine months. Some studies point to the widening of the Estacio as one of the first alterations in the geomorphology of the lagoon that has contributed to its rapid deterioration since the 80s of the past century.

11 These two factors have been studied and found responsible for the increase of nutrients, chemical substances (like phosphorus), and pollutants (from the mining activity developed in the southwest part of the Mar Menor) entering the lagoon and contributing to the process of eutrophication of its waters.

12 Thus, according to the 2016 scientific report (León & Bellido, 2016), one of the best measures to prevent that the big quantities of nutrients from entering the lagoon (which is one of the main causes of the process of eutrophication at the basis of the overall deterioration of the ecological equilibrium of the lagoon) are coastal wetlands, such as the ones that are characteristic of San Pedro del Pinatar and Marchamalo. In this sense, the preservation of these areas is not only motivated by their high biodiversity and cultural value but also by its further beneficial role in diminishing the impact of one of the main causes of the Mar Menor deterioration. In this sense, the project of recuperation of the wetland of the Carmolí area (in Los Urrutias) is conceived as a measure to diminish the

entry of nutrients via Albujón dry river bed (one of the main sources of nutrients entry into the lagoon) (León & Bellido, 2016).

13 The Saline of Marchamalo has been declared BIC (Bien de Interés Cultural) by Región de Murcia (BORM, 111, 2023).

14 The cultural significance of this activity and its recognition as a cultural landscape has been recently examined in (Sabaté Bel, 2020).

15 The development of a vocabulary which reflects some aesthetic awareness is one of the criteria that Berque proposed to identify a sensitivity to landscape and landscape appreciation (Berque, 1995).

16 An insightful study and reflection on how language reflects forms of perceiving and appreciating environment is (Macfarlane, 2016).

17 Still, as (Huxster, 2022) has shown by paying attention to the factors that may be relevant for assessing this influence in the context of current USA politics, there are many conditioning factors that modulate this impact and that overturn some of its expected outcomes.

18 I think this requirement is common to all appreciative practices. In art appreciation, for example, it is a common assumption that one has to possess some familiarity with the relevant aspects of an artistic practice to be able to appreciate a work within that practice. Similarly, there is more or less consensus in Environmental Aesthetics that proper appreciation of environments requires, if not full knowledge of the natural and culturally relevant facts about that environment, some acquaintance with the most significant facts. In this sense, aesthetic appreciation is necessarily framed by certain knowledge, thoughts, patterns of attention, normative expectations, etc.

19 For a view which embraces a more relativistic understanding of the aesthetic appreciation of human environments see (Maskit, 2014). For a view which embraces pluralism about aesthetic value in relation to the appreciation of cultural landscapes see (Cooper et al., 2016).

20 Although, as indicated earlier, this environment has experienced an accelerated deterioration since the 70s, the convergence of the factors that have put the ecological sustainability of the lagoon under critical conditions produced some alarming phenomena which were not only eloquently indicating that the critical state of the lagoon, but also fostering the population's sense of losing their environment.

References

Albrecht, G. (2005). Solastalgia: A New Concept in Human Health and Identity. *Philosophy Activism Nature*, *3*, 41–55.

Albrecht, G. A. (2020). Negating Solastalgia: An Emotional Revolution from the Anthropocene to the Symbiocene. *American Imago*, *77*(1), 9–30. https://doi.org/10.1353/aim.2020.0001.

Albrecht, G., Sartore, G.-M., Connor, L., Higginbotham, N., Freeman, S., Kelly, B., Stain, H., Tonna, A., & Pollard, G. (2007). Solastalgia: The Distress Caused by Environmental Change. *Australasian Psychiatry*, *15*(1_suppl), S95–S98. https://doi.org/10.1080/10398560701701288.

Arntzen, S., & Brady, E. (Eds.). (2008). *Humans in the Land: The Ethics and Aesthetics of the Cultural Landscape*. Unipub.

Ballesteros Pelegrín, G. A., Belmonte Serrato, F., & Sánchez-Sánchez, M. A. (2018). Las encañizadas del Mar Menor (Murcia, SE España): Ejemplo de recuperación de un modelo de pesca sostenible y respetuoso con la biodiversidad marina y el paisaje. *Cuadernos Geográficos*, *57*(3), 222–242. https://doi.org/10.30827/cuadgeo.v57i3.5986.

Berleant, A. (2010). *Sensibility and Sense: The Aesthetic Transformation of the Human World*. Imprint academic.

Berleant, A., & Carlson, A. (2007). *The Aesthetics of Human Environments*. Broadview Press.
Berque, A. (1995). *Les raisons du paysage: De la Chine antique aux environnements de synthèse*. Hazan.
Berque, A. (1998). Landscape and Immanence. *Thesis Eleven, 54*(1), 106–116. https://doi.org/10.1177/0725513698054000010.
BORM, 111. (2023). *Decreto n.o 136/2023, de 11 de mayo de 2023, por el que se declara Bien de Interés Cultural con categoría de sitio histórico, las Salinas de Marchamalo de Cabo de Palos, en el término municipal de Cartagena.* https://www.borm.es/#/home/anuncio/16-05-2023/2979.
Brady, E. (2006). The Aesthetics of Agricultural Landscapes and the Relationship between Humans and Nature. *Ethics, Place & Environment, 9*(1), 1–19. https://doi.org/10.1080/13668790500518024.
Brady, E. (2008). La estética de los entornos modificados. In Francisca Pérez Carreño (Ed.), *Estética del entorno: Obra pública y paisaje (2007-2008)* (pp. 4–22). Ministerio de Obras Públicas/CEDEX-CEHOPU.
Brady, E. (2023). Aesthetic Value as a Relational Value. *The Journal of Aesthetics and Art Criticism, 81*(1), 81–82. https://doi.org/10.1093/jaac/kpac066.
Budd, M. (2002). *The Aesthetic Appreciation of Nature: Essays on the Aesthetics of Nature*. Clarendon Press; Oxford University Press.
Cooper, D. E. (2008). *A Philosophy of Gardens*. Oxford University Press.
Cooper, N., Brady, E., Steen, H., & Bryce, R. (2016). Aesthetic and Spiritual Values of Ecosystems: Recognising the Ontological and Axiological Plurality of Cultural Ecosystem 'Services'. *Ecosystem Services, 21*, 218–229. https://doi.org/10.1016/j.ecoser.2016.07.014.
Danto, A. C. (1996). *The Transfiguration of the Commonplace: A Philosophy of Art* (Seventh printing). Harvard University Press.
Haapala, A. (1999). Aesthetics, Ethics, and the Meaning of Place. *Filozofski Vestnik, 20*(2), 253–264.
Haapala, A. (2005). On the Aesthetics of the Everyday: Familiarity, Strangeness, and the Meaning of Place. In *The Aesthetics of Everyday Life* (pp. 39–55). Columbia University Press. https://cup.columbia.edu/book/the-aesthetics-of-everyday-life/9780231135023.
Huxster, J. (2022). Influence of Environmental Movements on Public Opinion and Attitudes: Do People's Movements Move the People? In Maria Grasso & Marco Giugni (Eds.), *The Routledge Handbook of Environmental Movements* (pp. 472–497). Routledge.
Kant, I. (2000). *Critique of the Power of Judgment* (P. Guyer, Ed.; E. Matthews, Trans.; 1st ed.). Cambridge University Press. https://doi.org/10.1017/CBO9780511804656.
Kibebew, K. (2021). *Aesthetic Appreciation of Landscapes Contributing for National Identity. Case of Ethiopian Landscape.* Swedish University of Agricultural Sciences.
León, V. M., & Bellido, J. M. (2016). *Mar Menor: Una laguna singular y sensible: evaluación científica de su estado.* Instituto Español de Oceanografía Ministerio de Economía y Competitividad.
Levinson, Jerrold. (2006). Aesthetic Contextualism. In *Aesthetic Pursuits. Essays in Philosophy of Art* (pp. 17–27). Oxford University Press.
López-Martínez, G., & Espeso-Molinero, P. (2020). Pesca artesanal, patrimonio cultural y educación social.: El pescador murciano como transmisor cultural. *Revista Murciana de Antropología, 27*, 11–32. https://doi.org/10.6018/rmu.427471.
Macfarlane, R. (2016). *Landmarks* (Published with an additional glossary). Penguin Books.
Maes, H. (2022). Existential Aesthetics. *The Journal of Aesthetics and Art Criticism, 80*(3), 265–275. https://doi.org/10.1093/jaac/kpac018.

Maes, H. (2023). Aesthetic Melancholy. *Contemporary Aesthetics, 21*. https://contempaesthetics.org/2023/06/20/aesthetic-melancholy/.

Maskit, J. (2014). On Universalism and Cultural Historicism in Environmental Aesthetics. In J. Maskit (Ed.), *Environmental Aesthetics* (pp. 41–58). Fordham University Press. https://doi.org/10.5422/fordham/9780823254491.003.0004.

Nomikos, A. (2018). Place Matters. *The Journal of Aesthetics and Art Criticism, 76*(4), 453–462. https://doi.org/10.1111/jaac.12598.

Pechanec, V., Kilianová, H., Brus, J., & Machar, I. (2020). Analysis of the Aesthetic Value of the Cultural Landscape in the Territory of the Olomouc Archdiocese. *Public Recreation and Landscape Protection - With Sense Hand in Hand? Conference 2020*. Public Recreation and Landscape Protection - With Sense Hand in Hand? Conference 2020. https://www.scopus.com/inward/record.uri?eid=2-s2.0-85087274342&partnerID=40&md5=80f88754cd707b4fa45c5b3fdb98c4d4.

Ramsay, J., & Han, F. (2012). Space Is Not Nothing: Heritage Aesthetics and the Struggle for Space. *ICOMOS 17th General Assembly*, 119–124. https://openarchive.icomos.org/id/eprint/1128/1/I-2-Article4_Ramsay_Han.pdf.

Roger, A. (1978). *Nus et paysages: Essai sur la fonction de l'art*. Aubier.

Ruiz, J. M., Albentosa, M., Aldeguer, B., Álvarez-Rogel, J., Antón, J., Belando, M. D., Bernardeau, J., Campillo, J. A., Domínguez, J. F., Ferrera, I., Fraile-Nuez, E., García, R., Gómez-Ballesteros, M., Gómez, F., González-Barberá, G., Gómez-Jakobsen, F., León, V. M., López-Pascual, C., Marín-Guirao, L., Martínez-Gómez, C., Mercado, J. M., Nebot, E., Ramos, A., Rubio, E., Santos, J., Santos, F., Vázquez-Luis, M., Yebra, L. (2020). *Informe de evolución y estado actual del Mar Menor en relación al proceso de eutrofización y sus causas* (p. 165) [Informe de asesoramiento técnico del Instituto Español de Oceanografía]. Instituto Español de Oceoanografía (IEO).

Sabaté Bel, J. (2020). Las salinas, algunos retos como paisaje cultural = Saltworks, some challenges as a cultural landscape. *Cuadernos de Investigación Urbanística, 129*, 38–46. https://doi.org/10.20868/ciur.2020.129.4403.

Saito, Y. (2022). *Aesthetics of Care: Practice in Everyday* Life. London: Bloomsbury Academic.

Saito, Y. (2017). *Aesthetics of the Familiar: Everyday Life and World-Making*. Oxford: Oxford University Press.

Saito, Y. (2010). *Everyday Aesthetics*. Oxford: Oxford University Press.

Schama, S. (1996). *Landscape and Memory* (1. Vintage Books ed.). Vintage Books.

Taylor, K., St. Clair, A., & Mitchell, N. (Eds.). (2015). *Conserving Cultural Landscapes: Challenges and New Directions*. Routledge.

WHC. (n.d.). *The Operational Guidelines for the Implementation of the World Heritage Convention*. UNESCO World Heritage Centre. Retrieved 13 May 2024, from https://whc.unesco.org/en/guidelines/.

7 Monstrous lagoon

Creatures of art and science in the Mar Menor[1]

Maria Ptqk[2]

> Remember that *monsters* have the same root as *to demonstrate*; monsters signify.
>
> (Haraway, 2004, p. 117)

Monsters and ecosystems

In the autumn of 2019, a mass die-off of fish occurred in the Mar Menor after days of heavy rainfall. It was as if all the prophecies were fulfilled. It was foretold in 2016 by the so-called 'green soup,' the popular name given to the process of eutrophication, which happens when there is an overproduction of algae and phytoplankton due to agricultural fertilizers. Back then, waters had also turned fluorescent green, making an eerie sight. However, this new chapter reached a distinct level, with tons of dead fish washed up on the shores. 'The crabs tried to escape through the rocks, the shrimps piled on top of one another, practically making a paste, the eels jumped, the water boiled,' told a neighbor in the wake of the disaster. The news about the collapse of the marine ecosystem, which had been warned about by scientists and ecologists for decades, took on a cinematographic dimension, creating a strongly tragic horror-film-like atmosphere, as if it were the chronicle of a biblical plague. After decades of warnings, it seemed that the monster of the lake had been awakened.

This event signaled a shift in the dominant narrative, bringing the environmental health of the territory into the public arena – When is the balance broken? For what reasons? How is it measure? How does it get repaired? Who are held responsible? Eventually, the image of the Mar Menor as a dream holiday destination, as the 'paradise between two seas' immortalized in many songs and movies, was thrown into a crisis. However, the most relevant aspect of the incident was not the replacement of the Paradise narrative by the Hell narrative, but the putting into dialogue of both of them. This way, a more complex and systemic picture of Mar Menor emerged, more in line with the definition of real territory, that is, a living, multidimensional and dynamic environment, constantly modified by a myriad of entangled factors. The Mar Menor ceases to be a postcard and instead emerges in the collective consciousness as an ecosystem, just the opposite of a still picture.

DOI: 10.4324/9781003489078-8

The use of the figure of the monster is a way to point out that ecosystems are inherently monstrous realities, because they are multifaceted, uncontrollable, unpredictable and can only be partially approached. Like the underwater creatures of science fiction, of which humans can only have a partial glimpse (Prouteau, 2014) – a piece of the tail, a flank, a part of their head – ecosystems too are difficult to comprehend in their wholeness. A profusion of data and technologies allows us to identify only some of their components, such as chemical modifications in the atmosphere, on land or in water; or microscopic presences at the cellular or molecular level. But the entirety of the creature, the ecosystem itself, remains latent, phantasmagoric, unattainable. This condition of 'haunted landscapes' (Gould & Sermon, 2020) makes ecosystems legible, not only through the lens of science but also through perception, invocation and sensitivity.

Art-science: a fabulous creature

This kind of fabulous condition, of mythical creature, is also found in practices identified under the heading of art-science, such as those developed in the context of *Reset Mar Menor*. In fact, it is quite relevant that artists Clara Boj and Diego Díaz had created an immersive reality artwork before this long research project began, starring precisely a sea monster. In their art project, called *ISBE: Mar Menor Research*, they bred an artificial life organism under the observation of the Instituto para el Estudio de los Enigmas Biológicos [ISBE, Institute for the Study of Biological Enigmas]. The creature changes its appearance depending on the data it is fed: temperature, water composition, pollution, hotel occupancy, etc. Biological and social data are combined to produce a technologically alive naturecultural entity that allows itself to be seen in different sites of the lagoon. The project, designed both as an artwork and as a research artifact, adequately synthesizes the multidisciplinary strategies that ecosystems require.

The category of art-science, considered as a non-field even by its own practitioners, comprises an array of artistic strategies usually described as hybrid, always in search of a (maybe unnecessary) definition (Bureaud, 2016). Understood as experimental exercises, they test languages and methods; when viewed as speculative practices, they expand knowledge and mental representations; and as artistic devices, they work with the symbolic and sensitive, generating a kind of meaning not always translatable in discursive terms. They produce a sometimes clearly inter- or transdisciplinary knowledge, which most of the time is rather extra-disciplinary (Holmes, 2007). For this reason, the strange alliance between arts and sciences is suitable for nonconventional research, such as investigations increasingly conducted in the field of ecology. This kind of research requires methodological invention to call into question some accepted notions of knowledge and expertise, sometimes even refusing to accept the status of 'object of study,' in order to open up to the possibility that this so-called 'object' might be a subject with agency and trajectory (Latour, 2015). This is why being described as a monster suits it so well, following Donna Haraway's description of monsters as inappropriate/d others, the ones that, precisely because of being strangers, generate meanings differently (Haraway, 2004). For monsters, as Haraway says, signify.

Therefore, monstrosity works here on different levels. First, it makes reference to the notion of Mar Menor as neither a landscape nor a territory, but as a living system composed of communities of beings that share the same physical environment, and whose face changes according to inner entanglements. Second, it refers to the kind of artistic practices developed within the context of the *Reset Mar Menor* project. Even though some of them are not recognized as belonging to the art-science category, they all include diverse degrees of practice-based research that allow them to move between different fields of knowledge and search for their codes of belonging in the no-one's land of extra-disciplinarity. The following is a selection of some of these projects, the ones whose characteristics help us approach art-science, that kind of strange, probably imaginary genre.

Techniques of objectivity

We will begin this journey with the map and the documentary, two techniques that, although not considered specific to the art-science genre, are common in the dissemination of scientific knowledge and have a significant genealogy in the arts field. Both of them unify perspectives, establish the boundaries of a territory or issue and produce an appearance of objectivity as an exact mirror of reality, to the extent that they can be taken as reality itself.

Cartography is the classic technique used to represent the territory, flattening and normalizing it. It defines its shape and composition, determining what is contained within and what remains outside. Everything that shapes the subjective and heterogeneous experience of the territory – everyday stories, memories, the diversity of sights, the contradictions – is traditionally kept beyond the maps. For this reason, probably the most important action of the *Reset Mar Menor* project was the creation of a collective critical cartography group. *Líneas de costa* [Coastlines] is a collection of unfinished, tentative, and personal maps that provide a cartography of the saltwater lagoon from various perspectives, including migratory, social, labor and environmental dimensions. Participants shared their experiences around a table in the shape of the lagoon, allowing for the emergence of a dense and polyphonic identity out of the territory. New maps were superimposed over the geographic map – which outlines coastlines or inner islands – and over the political map – which locates towns or the road network – , thereby expanding the imaginary map. 'The mud baths in Las Punticas, the traffic jams in La Manga, the jellyfish, the boat trip to La Perdiguera, the nitrates leaking in Rambla del Beal, the golf courses and the airport…'.[3] 'These are speaking maps', said a visitor to the exhibition where the results were first shown.[4]

The use of a critical collective cartography is common in artistic research and in militant or action-research. Its ability to embrace various voices multiplies perspectives on territories, while calling into question the idea of the cartographic technique as a way of producing univocal and closed narratives about landscapes. In critical cartography, maps are conceived as creative acts that, by enabling multiple approaches to the same issue, are useful to 'bring people together and find oneself in a shared reality,' in the words of Brian Holmes. 'Maps,' he says,

> work with imaginary and hidden elements, they have to do with the sky above us and with the underground. Somehow, they have to do with utopia, with our vision of the future and with the problems we see in the present.
> (Holmes, 2006)

On the other hand, the documentary genre, whether scientific or anthropological, has also been frequently revisited from visual studies and artistic research, with the aim of analyzing, or sometimes subverting, the methods used to transform objectivity into an artifact. Shots, editing, voiceover and musical accompaniment are among the techniques that contribute to constructing the gaze in a way that not only appears as truthful and reliable, but also as the only possible perspective. The figure of the expert plays a key role in addressing that process, being the voice of authority that provides the technical or scientific knowledge, that is, the objective knowledge required to interpret what is being seen. This is precisely one of the most interesting aspects of the documentary *Memorias del Mar Menor: la pesca en un paisaje en crisis* [Memories of the Mar Menor: Fishing in a Landscape in Crisis] (2020), by Julio Daniel Suárez: the role it assigns to the fishermen of the Cofradía de San Pedro del Pinatar, the fishing cooperative. In front of the camera, they not only recount the difficulties and satisfaction that come from their profession, the attractiveness of the sea and the difficulties of generational turnover, but they also accurately describe the ecological conditions of the lagoon, as keen observers of the environment after decades of spreading their nets on the spot. For instance, their experience has taught them that, although eutrophication initially increases the fish population – due to an increase in nutrients in the water – , this apparent abundance is only a symptom of imbalances that will eventually damage fisheries. They also know that the variety and taste of fish are not the same as they were fifty years ago and that, although the causes are many, the opening of the Canal de la Estació in 1973 marked a turning point, as the canal was dredged to facilitate access to the marina, thereby reducing its salinity and altering the balance between species. As with the collective cartographies, the men and women of the sea depict a rich and complex picture of the ecosystem. By doing so, they assert themselves as experts in that particular territory, in an exercise of knowledge that is embodied in their own lives and traversed by their subjectivity. This kind of knowledge is endowed with a truth which is absent in the abstract knowledge of the experts, who lack an emotional or biographical connection to what they observed.

Citizen ecology

Creating other places of enunciation, showing that there are other ways of acquiring knowledge and other forms of legitimacy in the production of discourse, is a recurring strategy in art-science practices. A significant portion of practitioners under this category aims to 'bring science out of the lab and stage it in the public domain, giving people direct experience of common scientific processes' (Triscott, 2009). Convinced that 'science doesn't just think about the world,' but 'it makes the world and then remakes it,' they consider scientific knowledge to be a political space, and as such, it should include public participation. Despite being regarded as a contemporary phenomenon, the genealogy of citizen science actually dates back to the beginnings of modern science, whose development would not have been possible without the input from amateur communities. The large collections of Natural History Museums, as well as empirical knowledge in natural sciences – such as botany, entomology, mycology and zoology – , have been nourished by the specimens collection and data-gathering carried out by groups of amateurs, many of whom women, without academic training but with systematic and sometimes obsessive dedication (Zhong-Mengual, 2021).

Various projects in *Reset Mar Menor* are inspired by this amateur spirit, which is now part of the DIY and DIWO (Do-It-Yourself and Do-It-With-Others) body of practices inherent in the genesis of art-science. One of them is the collective fanzine *Bajo fondos submarinos, praderas de posidonias y medusas* [Undersea Depths, Posidonia Seagrass Beds, and Jellyfish], led by the artist Gelen Jeleton, that brings together images, drawings, collages, texts and photographs to compose a coral and heterodox picture of the lagoon's aquatic ecosystem. Another one is the workshop *Microuniverses* [Microuniverses], facilitated by Hamilton Mestizo, who introduces participants to the building of homemade microscopes to analyze water, soil, air or vegetation samples. This artist understands his practice as a form of activism. His work is framed within biohacking or biotech activism, a movement driven by radical curiosity and a desire for accessibility that advocates for the large-scale availability of biotechnology knowledge and tools. 'We recognise that some of the greatest cultural and technological advances have emerged from people's bedrooms and are therefore committed to transferring the high-tech life sciences to the bedroom biotechnician,' claim Heath Bunting and Natalie Jeremijenko, authors of the seminal magazine *The Biotech Hobbyist*. Would it be possible to teach the residents of Mar Menor how to collect and analyze data on water quality? Would a community of neighborhood hackers responsible for monitoring levels of heavy metals, nitrates, or oxygen be feasible?

Within the extensive repertoire of amateur practices, one stands out for its simplicity and efficiency. It does not require any sophisticated techniques, aside from facilitating analytical observation and functioning as a tool for registering and communicating, characteristics that have made it the most widely employed method by scientists throughout history. Even today, in the digital age, the basic drawing technique – paper, pencil, and freehand – still remains a key element in the making of knowledge, especially in biology and natural sciences. Many great naturalists have also been excellent sketchers, from Linnaeus or Alexander von Humboldt to Ernst Haeckel and Darwin. A tradition that continues to this day, such as in the work of botanist Francis Hallé, whose impressive tree drawings are displayed in exhibitions and artistic publications, and which attests to the frequently overlooked historical connection between scientific research and artistic techniques. Notes, sketches or visual diagrams that allow us to capture what is observed while also helping us think about it, highlighting the connection between the sense of sight, the movement of the hands and the functioning of the brain (Maitz, 2014).

This sketch approach or draft is included in both *Dibujando el Antropoceno* [Drawing the Anthropocene] (2018) and *Dibujando el Capitaloceno* [Drawing the Capitalocene] (2019), two tours of the old mining area in Sierra de Cartagena. In them, participants create field notebooks to record information about what they learn and perceive from the environment, with the help of scientific and artistic professionals. The drawings depict a landscape devastated by toxic waste, with tourist apartments and luxury resorts, alongside stratigraphic schemes and notes on mineral extraction techniques. Among all these records, which contrast with each other because they belong to very different visual and knowledge registers, personal insights slip through, such as how to connect with the recent past of mining, how to tell stories of the rock that go back billions of years or how to face challenges with different outcomes depending on the time frame used.

Methodological imagination

This search for liminal knowledge between the poetical and the empirical that emerges from practices inspired by amateur culture can also be found in the methodological imagination, which is a point of collision but also a creative encounter between art and science. One of these generative frictions comes from art using marginal techniques that have become controversial or outdated for not following scientific protocols.

The audiovisual installation *Cartografías de la metamorfosis* (2020) [Cartographies of Metamorphosis], by Pedro Ortuño, uses images generated by the sensitive crystallization method to compare fruits and vegetables from biodynamic agriculture and industrial agriculture. This technique was invented in the 1920s with the aim of determining the intrinsic quality of organic material by the analysis of the graphics generated by its crystallization with copper chloride salts. In his project, Ortuño shows that vegetables from biodynamic agriculture, forerunner of permaculture, create harmonious patterns resembling mandalas, whereas those from industrial framing generate narrow and unstructured shapes. Within the context of this artwork, sensitive crystallization visually captures the extreme violence against living beings entailed in intensive agriculture, blurring the boundaries between scientific visualization and artistic creation, between objective and evocative images. In fact, his inventor was the biochemist Erhenfried Pfeifer, an early ecologist who advocated for agriculture based on natural soil cycles. He was among the first scientists to speak out against the use of pesticides, such as DDT (Dichlorodiphenyltrichloroethane), whose health risks Rachel Carson would later write about in *Silent Spring*.

In the face of the hyper-formalization of scientific protocols, artistic research projects such as Susana Cámara Leret's *Tierra de diatomeas* [Land of diatoms] undertake the challenging task of imagining their own working methodologies with protocols, languages and ways of doing inherited from other disciplines, which have been transformed or invented *ad hoc*. Within the context of *Reset Mar Menor*, this project took the form of a workshop in which participants were introduced to basic techniques, such as sample collection or microscopic analysis, while they also approached knowledge regimes as distant from science as traditional architecture or the arts of dowsing. In this project, methods and conceptual innovation go hand in hand. Novel methodologies are imperative in seeking an unusual perspective on the subject matter under study. On the one hand, the project investigates the movement and behavior of diatoms, an unprecedented approach in conventional diatomology. On the other hand, it explores the potential of diatoms as a figure of thought, as a conceptual tool for reflecting on the relationships between humans and their surrounding environment, the aquatic origin of life or the grand geological narratives of evolution. The singularity of these deeply transdisciplinary approaches, which fully exploit the cross-language at the intersections of art and science, is that they create working environments where questions arise that could not be formulated from any single discipline alone. These shared questions are both common and unknown to all of them, and they compel the search not only for new techniques and methods, but also for new vocabularies, metaphors and mental images.

It is likely not a coincidence that the microorganisms chosen to stimulate research imagination are diatoms. These unicellular microscopic algae, present in all aquatic environments and used as bioindicators of ecosystem health, are old acquaintances in the art field. They are the main characters of probably one of the most important works of scientific illustration in the history of art, the celebrated *Kunstformen der Natur* (Art Forms in Nature) by biologist, philosopher and artist Ernst Haeckel. His iconic drawings of radiolarians, jellyfish, sea anemones and diatoms, which introduced the exuberance of the marine microcosm to early 20th-century society, also inspired the modernist taste for natural forms, a strong influence in Art Nouveau design and architecture. But that is not all. While Haeckel was drawing diatoms trying to accurately capture their details, new words were taking shape in his mind. Words that have now become established in biology were neologisms at the time, some of them profoundly evocative. *Phylum* to refer to the taxonomic category between kingdom and class; 'ontogeny' to denote the branch of biology responsible for studying the life development of an organism, from fertilization to death; and 'ecology', the combination between *oikos* (the Greek term for home) and *logos* (knowledge), to name the new science with a holistic vocation that was aim to understand the environment as a common home (*oikos* or household).

Post-naturalist media

Imagery is another meeting point between art and science, with numerous manifestations in biology. In a society accustomed to all sorts of devices to create images with an appearance of veracity, where everything is mediality in terms of technology-mediated representation, the concept of 'natural' is diluted. As a result, it has become harder to think about the status of imagery, as many artists and philosophers interested in the codes of visuality and its 'regimens of truth' have done. To which visual category does an image of a cell seen under a microscope belong to? And a 3d rendering of a virus obtained through cryo-electron tomography? Is there any difference between a technical image for scientific purposes, which reproduces a natural element *as neutrally as possible*, and a media image constructed through the accumulation of layers of interfaces, which gives the image a genealogy and a point of view? According to art historian Gabriele Werner, '[t]he technical image is to be taken seriously because of its formal and media-based functional property, as if it were an autonomous artwork.' This is because '[i]t shapes communication and produces knowledge such as devotional images or political pamphlets' (Maitz, 2014).

Contemporary visualization technologies used in biology, geography or environmental sciences produce a type of technical imagery that could be designated as post-naturalistic, for lack of a better term. An example of this is generative graphics, such as those created by the artist and programmer Alba G. Corral, who facilitated a workshop called *Mar de datos* [Ocean of Data], in which she used the Processing programming language to relate data extracted from water to changes in its color. Studies on water color allow for the determination of variables, like the presence of phytoplankton and sediments, temperature or salinity, producing unprecedented and highly expressive visualizations that are both tools for biochemical analysis and creative representations of the marine environment.[5] Images that produce meaning behind the appearance of a technical instrument: through their mediation, water emerges as a living being, animated by invisible entities, connected to and conditioned by its environment.

Many media art projects with an interest in an ecological perspective share this desire to reveal other dimensions of reality. They portray natural environments as mysterious creatures – once again the monster – that conceal more than they reveal, and whose working cannot be understood only through the mechanics of science. The work *Out of Sight, Out of Mind* (2020) by Paul Sermon, Jeremiah Ambrose and Charlotte Gould, created during the *Reset Mar Menor* production residencies, aims to explore this 'metaphysics of the landscape.'[6] It is filmed with a 360° camera and incorporates on-site video and audio recordings, as well as telepresence devices, that draw participants living the 3D experience into the immersive media itself. The installation offers an unexpected sight of the lagoon that reveals its absurd, ironic, poetic and unsettling latent identity. A fried egg escapes from the classic combo plate and swims through the sky over a sea of pink salt flats, avoiding collision with skyscrapers; the ruins of the old military base on Perdiguera Island catch fire, struck by projectiles of unknown origin; a mining cart hangs suspended in the void over barren land, likely one of the ravines where toxic waste was dumped into the lagoon; a conversation about shrimp fishing in a bar full of mirrors that resembles a painting from the Spanish Baroque era, in which characters have come to life. Inspired by Merleau-Ponty's phenomenology of perception and his studies on conscious formation, and by the exploration of senses by researchers such as Roy Ascott, this art research project expands the field of the sensible through technical mediation (Gould & Sermon,

2020). The imagined landscape overlaps with the empirical territory, forming a kind of unconscious of the place.

This perspective is shared by Laboratorio de Luz [Light Lab], the collective behind *Bajo la manga*[7] (2020), a media installation that reveals the sonic identity of the landscape, an often overlooked aspect of territories. It includes sound recordings from three different locations in La Manga – a site polluted by mining, another one contaminated by tourism, and a third one in the salt flats area, less affected by human intervention –, along with images of marine fauna and flora, and others generated by audio data processing. The final result is materialized in an audio-visual installation and a performance and offers a multiplicity of visual and sound references that demonstrate the extraordinary narrative potential of the Mar Menor.

New scientific narrative

Almost all the projects developed at *Reset Mar Menor* contribute to updating the scientific narrative with varying degrees of explicitness, in order to attune it with contemporary ecological consciousness. In recent years, not only has the environmental crisis become a top political concern, but also the idea gains ground among philosophers and anthropologists, biologists and environmentalists, that we are witnessing a paradigm shift in our relationship with the environment, motivated by the crisis of what many authors call the 'great divide' between nature and culture, humans and nonhumans. According to this binary vision of the world, which also underpins the separation between art and science, the creation of meaning lies on the human side, along with intelligence and intentionality – all agency, in the words of Bruno Latour (Latour, 2015). On the other side, there is nothing but inert matter.

Much of philosophy of science from the early decades of the 21st century, as well as some artistic trends inspired by the emergence of the so-called 'multispecies ethnography' (Kirksey & Helmreich, 2010, p. 545), has been devoted to deconstruct this 'disenchanted vision of the world.' The characteristic feature of this field of knowledge – or state of consciousness – is that '[c]reatures previously appearing on the margins of anthropology – as part of the landscape, as food for humans, as symbols – have been pressed into the foreground in recent ethnographies.' Not only animals, plants, fungi and microbes, but also geochemical cycles, geological strata and the atmosphere, which were previously confined to the category of the inert – the realm of *zoé* or 'bare life', as called by Giorgio Agamben – have begun to fall under the category of *bios*, with its politics and histories (Agamben, 1998). From this point of view, nature is no longer the background against which human history is projected, nor is it an object studied by science or admired by art. Nature acquires 'an active voice' (Plumwood, 2013) that challenges the conventions of the great divide.

Important in this context is the work of boredomresearch collective, a collaboration between British artists Vicky Isley and Paul Smith. During their residency at *Reset Mar Menor*, they developed the first part of *Anthrobiotica* (2020), a project that departs from algal blooms associated with eutrophication episodes. The artwork is a suggestive underwater fabulation in which bioluminescent communities of microorganisms become the protagonists of the plot. As a result of a sensitive use of generative images and digital animation techniques, along with a careful sound production, the lagoon emerges as an eerily vivid and fantastical underwater world. This seemingly paradoxical effect is characteristic of these new scientific narratives, which update the approach to the natural world within the genre of Wonder. On the one hand, the artwork transforms the 'green soup' into a less enigmatic, more realistic phenomenon resulting from the interaction between sediment-dwelling colonies of bacteria, microalgae, fungi and phytoplankton. On the other hand, it suggests that this invisible cosmos stirring beneath the feet of bathers is better seen through the eyes of fantasy. Through artistic research, the artwork evokes underwater life in a way that makes the concrete existence behind scientific data tangible, merging knowledge with sensory experience.

The transformative power of fiction is also effective when these fabulation exercises are detached from reality, even when they are blatantly absurd. This is suggested by *Relocating Training Program for* Pinna nobilis (2020), a research project by Christina Stadlbauer that was developed in the production residency. The result was a video tutorial for *Pinna nobilis*, common name the Noble Pen Shell, a large mollusc endemic to the Mediterranean Sea and in danger of extinction due to overfishing and the loss of water quality. The video tutorial, produced by the Institute for Relocation of Biodiversity (an imaginary agency), provided *Pinna nobilis* with all the necessary information to move its habitat towards a safer, more wholesome environment. The film was premiered underwater in the context of *Reset Mar Menor* to a few human visitors wearing snorkels and goggles, accompanied by a percussive sound piece specifically created for the Noble Pen Shell, able to hear sounds inaudible to human ear.

With the exception of the 'assisted relocation,' which is an actual measure proposed by conservation organizations to mitigate the impact of habitat loss, everything in this artwork might be read as a mockery of that new multispecies sensitivity, determined to treat nonhumans as if they had decision-making power. The proposal, however, has the opposite effect. 'The impossible – absurd – solution of the video tutorial for a mollusc and screening it underwater,' the author says, 'is intended as a way to emphasize the urgency of the situation we find ourselves in. (…) [H]umans trapped in an incomprehensible world that has abandoned rational devices and discursive thought' (Stadlbauer & Bartaku, 2020, p. 47). A world unable to appreciate the abilities and intrinsic value of other life forms, or to give efficient responses to stop the destruction of ecosystems and the biodiversity loss. The elements of *le théâtre de l'absurde*, along with parody and irony, are here at the service of 'making kin,' Donna Haraway's invitation to broaden our affective kinship bonds to include other life forms (Haraway, 2016).

Art and science to re-enchant the world

Understanding the act of imagination as a political action due to its ability to build possible worlds is a constant in artistic practices that explore different ways of relating with the environment. The imaginaries generated in *Reset Mar Menor* do not give any solution to soil pollution or the loss of biodiversity; nevertheless, they provide a myriad of elements to tell other stories. Entering and leaving the lagoon, and through fragmentary visions, they offer a variety of perspectives in which other possible stories can be identified. Whether they are scientific tales, new methodologies, popular science initiatives or transdisciplinary actions, the projects developed in this context are aligned with what some consider to be a historical process of science 're-culturization,' one that places science firmly in its social context, facing the challenges of its time (Lévy-Leblond & Malina, 2016). From this point of view, science is not the 'culture of no culture' (Haraway, 2018, p. 23), but a particular cultural form, a narrative practice that generates stories about the world.

According to the art historian Estelle Zhong Mengual, the ecological imaginary, informed by conceptual tools developed for the contemporary philosophy of the living, has the potential to transform the relationships between knowledge and sensitivity, making them feed on each other until they merge (Zhong-Mengual, 2021). If the great divide gave science the task of disenchanting the world by reducing it to a mere accumulation of mindless matter, and gave art the monopoly of re-enchanting the world through evocation and poetry, then the linkage between art and science serves precisely as a way to break down this distinction, in order to understand it as a 'new form of science (…) that encourages learnedness in natural science along with all the tools of the humanities and the arts.'[8] In short, a convergence field or a 'composition'[9] of myriad languages, technologies, protocols, methods, institutions, and more… Each one with its own genealogy, but all of them engaged in a common task, which is not to find the correct answers, but rather to improve our questions.

Notes

1 This text was originally published in the book *Reset Mar Menor. Laboratorio de imaginarios para un paisaje en crisis* (Boj, 2023) that collects the development of *Mar Menor Lab*, an artistic research under a transdisciplinary approach developed between 2017 and 2021 in the Mar Menor lagoon. The project proposed the construction of new narratives for the territory based on its current state of fluid space of coexistence and conflict, by the articulation of knowledge and the collaboration between artists and scientists. Under the guidance of researchers Clara Boj and Pedro Ortuño, and with the support of Fundación Daniel and Nina Carasso, more than 30 artists and scientists were involved in the project, which was promoted by the research groups Arte y Políticas de Identidad and EcosisCOST-Ecología y Ordenación de Ecosistemas Marinos Costeros of the University of Murcia. http://www.marmenorlab.org/.

2 María Ptqk is an independent curator, researcher and cultural manager specializing in the intersections between the arts and techno-scientific culture.

3 Presentation of the first Critical Cartography Conference at the Mar Menor, which took place at the University of Murcia Museum, in July 2018.

4 Exhibition *Lagunas. Proyectos y procesos de arte y ciencia en el Mar Menor*, held in Centro Cultural Puertas de Castilla (Murcia), from 9th to 27th May 2019.

5 As for example, the artistic research project 'La couleur de l'eau,' by artist Nicolas Floc'h in collaboration with marine-optics expert Hubert Loisel, presented at the exhibition *Paysages productifs*, at FRAC Provence-Alpes-Côte d'Azur, 2020–2021.

6 Paul Sermon in the exhibition *Lagunas. Proyectos y procesos de arte y ciencia en el Mar Menor*, see note 4.

7 Untranslatable pun meaning both 'up one's sleeve' and 'under the Manga' (del Mar Menor). [T.N.].

8 Anna Lowenhaupt Tsing, 'Art of inclusion, or How to love a mushroom,' *Manoa* 22, no. 2, 2015, 191–203.
9 Bruno Latour, 'Il n'y a pas de monde commun: il faut le composer,' *Multitudes* 45, 2011, pp. 38–41.

References

Agamben, G. (1998). *Homo Sacer: Sovereign Power and Bare Life* (D. Heller-Roazen, Trad.). Stanford University Press.

Boj, C. (2023). *Reset Mar Menor. Laboratorio de imaginarios para un paisaje en crisis*. Bartlebooth.

Bureaud, A. (2016). Plaidoyer pour non-champ. *Magazine des Cultures Digitales, Arts&Sciences*, (81). https://www.digitalmcd.com/art-science/.

Gould, C., & Sermon, P. (2020). The Immersive Environment as a Driver for Environmental Change; Addressing the Out of Sight, Out of Mind Impacts of the Anthropocene on the Mar Menor, Spain. *Virtual Creativity*, *10*(2), 141–161. https://doi.org/10.1386/vcr_00029_1.

Haraway, D. (2004). The Promises of Monsters: A Regenerative Politics for Inappropriate/d Others. In Haraway, D. (ed.), *The Haraway Reader*, 63–124. Routledge.

Haraway, D. (2016). *Staying with the Trouble: Making Kin in the Chthulucene*. Duke University Press Books.

Haraway, D. (2018). *Modest_Witness@Second_Millennium. FemaleMan_Meets_OncoMouse: Feminism and Technoscience* (T. N. Goodeve, Ed.; Second edition). Routledge.

Holmes, B. (2006). Luz de estrellas y secretos; una breve charla sobre mapas y sus usos. En P. Monsell Prado (Ed.), *Fadaiat: Libertad de movimiento + libertad de conocimiento* (1a. ed). Centro de Ediciones de la Diputación de Málaga, 62–63.

Holmes, B. (2007). Extradisciplinary Investigations. Towards a New Critique of Institutions. *Transversal*. https://transform.eipcp.net/transversal/0106/holmes/en.html.

Kirksey, S. E., & Helmreich, S. (2010). The Emergence of Multispecies Ethnography. *Cultural Anthropology*, *25*(4), 545–576. https://doi.org/10.1111/j.1548-1360.2010.01069.x.

Latour, B. (2015). *Face à Gaïa: Huit conférences sur le Nouveau Régime Climatique*. Les Empêcheurs de penser en rond / La Découverte.

Lévy-Leblond, J.-M., & Malina, R. (2016). Un bref dialogue sur deux systèmes du monde. *Magazine des Cultures Digitales, Arts&Sciences*, (81). https://www.digitalmcd.com/art-science/.

Maitz, P. (2014). *Visualisation of Evolution: Molecule/calculus*. Ambra Verlag.

Plumwood, V. (2009). Nature in the Active Voice. *Australian Humanities Review (46) May*, 113–129.

Prouteau, E. (2014). Qui sont ces serpents? De Jules Verne à Huang Yong Ping, histoires d'un ophidien mythique. *Revue 303. Arts, recherches, créations, Images de Jules Verne*, (134), 142–156.

Stadlbauer, C., & Bartaku. (2020). The (Im)possibility of Communicating with Other Species. *Performance Research*, *25*(5), 45–48. https://doi.org/10.1080/13528165.2020.1868840.

Triscott, N. (2009). Performative Science: The case of Critical Art Ensemble. En M. Chatzichristodoulou, J. Jefferies, & R. Zerihan (Eds.), *Interfaces of Performance*, 153–166. Routledge.

Zhong-Mengual, E. (2021). *Apprendre à voir: Le point de vue du vivant*. Actes Sud.

8 A citizen assembly for transitioning the Mar Menor?

Nicolás Gutiérrez, Alba Ballester and Violeta Cabello

Introduction

Citizen assemblies (CAs) have been popularized as an innovative form of deliberative governance capable of unblocking public action on complex social and environmental problems. Propelled by demands from new environmental movements like Fridays for Future and Extinction Rebellion, recent years have witnessed a "wave" of assemblies for climate change mitigation and adaptation in many countries across Europe (Boswell et al., 2023). As one of the different formats of deliberative mini-publics (Escobar & Elstub, 2017), CAs bring together a sample of 100–200 randomly selected citizens to learn, think and deliberate on a particular policy issue, ultimately producing recommendations for future policies. Whereas climate change has gained particular traction as a deliberative topic, CAs have been developed on a wide range of challenging topics like mental health, crime, drug use, rural sustainability, youth problems or the future of the European Union, to name just a few.

One of the success stories in the CAs literature is the deliberation over abortion in Ireland during 2016–2018 that led to a major shift in existing regulations toward its liberalization. This case evidenced the democratic potential of aligning CA processes and parliamentary discussions, both in time and contents (Suiter et al., 2022). Moreover, it helped translate the public debate on the topic across societal groups leading to a major sociocultural transformation. This case is particularly relevant because abortion was a polarizing topic in Ireland and politicians had been unable to move legislation forward for over 30 years. After listening and learning from experts and actors supporting both sides of the problem, citizens proposed to remove constitutional articles that prevented legal abortion and this proposal was voted and approved in a national referendum.

Despite the great scientific consensus, climate change is also a polarizing policy issue in some countries (Karakas & Mitra, 2020). However, climate CAs do not often treat climate change as an issue with different positions that need to be listened to and accommodated. Rather, they are envisioned as a way to support campaigns for climate action, assuming participants are convinced of the need for such action (Crosby, 2021). Efforts in climate CAs tend to focus on the daunting task of unraveling the complexity of the topic, the many dimensions, sectors

DOI: 10.4324/9781003489078-9

and scales involved for citizens to gain a minimum climate-literacy in order to deliberate and produce recommendations. Thereby, it is expected to bridge the gap between an international climate urgency agenda and daily social practices and to overcome the political paralysis toward climate action (Willis et al., 2022). The extent to which this transformative impact gets catalyzed by climate CAs is still under examination (Boswell et al., 2023).

Among the many challenges ahead for CAs as democratic innovations for environmental governance, this chapter reflects on those related to their organization in strongly contested environments such as the Mar Menor (see KNOCA, 2023a). In the last years, sociopolitical tension has risen as the lagoon shows signals of severe eutrophication and public administrations seem incapable of taking effective action (Cabello & Brugnach, 2023). The stagnation of the situation motivated calls for a CAs among environmental organizations. For this reason, we were invited to run a one-day epistemic experiment during the interdisciplinary training that weaves the chapters of this book. The purpose of the exercise was to experience and reflect upon what could happen in a real deliberative CAs when a diversity of views sit together to define future actions. We simulated such a diversity of positions by giving human and more-than-human roles to half of the participants. Listening to other living beings and future generations is another emerging challenge in climate CAs (KNOCA, 2023b), of particular relevance to the Mar Menor in the context of its recent legal personhood.

The goal of this chapter is to contribute to learning about how to sit in “the fire” of conflictive interactions (Mindell, 2014) within CAs. In addition, the presented experimental simulation may support educational and research efforts that advance the boundaries of environmental CAs (Cherry et al., 2021). For this purpose, we develop a systematization of the experience both of the methodology and of the conversations engendered. Systematization of experiences is a widespread methodological approach in participatory action research and critical pedagogies (Jara, 2012; Mejía, 2012).

The Mar Menor as an environmental conflict

Environmental conflicts can be broadly defined as social conflicts related to the environment (Scheidel et al., 2020). They may overlap with other dimensions of social conflicts such as class, gender, ethnicity or identity. Despite early warnings by scientists, the Mar Menor eutrophication became a social conflict after the “green soup” of 2016. The first algae bloom marks an important shift in the history of the lagoon as new environmental movements like Pacto por el Mar Menor emerged to claim solutions (Zuluaga-Guerra & Cabello, 2023). They actively learnt and disseminated information about eutrophication, pointing at intensive agriculture as the main source of nutrient enrichment. In turn, the agricultural sector organized a counter-narrative to dispute the centrality of agricultural responsibility (Cabello & Brugnach, 2023). The public dispute intensified after in 2018, when the Scientific and the Participation Committees created by the Regional Government of Murcia tore apart. That same year, the agricultural sector created a foundation that has actively contributed to escalating tensions by spreading misinformation that denies the connection between farming practices and eutrophication of the lagoon.

Beyond these two dominant narratives on whether agriculture is or not the main driver of eutrophication, the Mar Menor constellation of actors affecting or affected by the lagoon's crisis is wide and complex. Previous stakeholder analyzes identified the following sectors: citizen platforms, neighborhood associations and NGOs; academy; fishing; construction and real estate; tourism; agriculture; public administrations; and local population (Guaita-García et al., 2022; Boix-Fayos et al., 2023). Most of these sectors are multilevel with local organizations connected to regional, national and/or international actors. This is particularly relevant for the agricultural sector, a transnational privately governed and deeply asymmetric network connecting large European retailers to agro-export business, farmers, farming unions, irrigation unions and workers, mostly migrants, on fields, pig farms and postharvest industries (Pedreño et al., 2022).

The public governance of eutrophication in the Mar Menor is also scattered throughout levels of governance: local municipalities in the basin (Cartagena, Torre Pacheco, San Javier, San Pedro, Los Alcázares, La Unión, Murcia, Fuente Álamo), in charge of urban planning and wastewater management; the regional Government of Murcia, with competences over land planning, environmental protection and agricultural development, and a specific directorate for the Mar Menor. The Spanish Government is responsible for water management through the Segura Water District and with a dedicated office within the Ministry for Ecological Transition. Over the past few years, the presence of opposite political colors in different administrations has significantly hindered the development of cooperative plans and regulations. This is one of the key reasons why achieving a genuine CA in the region remains difficult to envision.

It is noteworthy that no new formal space for multi-stakeholder dialogue has been created since 2018. The new Law 19/2022 on the Mar Menor Legal Personhood foresees several participatory committees, one of which will be soon operating. Additionally, Law 3/2020 for the Mar Menor protection and recovery from the Regional Government of Murcia mandates the creation of several committees, including a Council for the Mar Menor.

Designing a simulation of CAs for the Mar Menor

Our methodological proposal stems from experiential learning and action-oriented pedagogies, such as those of Paulo Freire, that subvert the roles of teacher and student to engage in collective experimentation and reflection as avenues for whole-person learning (Kolb & Kolb, 2005; Sipos et al., 2008; Craps & Brugnach, 2021). We designed a simulation of a regional CA tasked with producing recommendations under the following mandate: "For a less eutrophic Mar Menor, what changes are necessary?".

The mandate of a CA is the overarching question that guides the whole deliberative exercise and the final production of recommendations. The proposed mandate aligned with the latest report from the Spanish Institute of Oceanography on the situation of eutrophication in the Mar Menor and its causes (Ruiz et al., 2020). The report signals that (a) the Mar Menor has shifted from an oligotrophic to an eutrophic state defined by nutrient overenrichment of its waters; (b) nutrients come from a different point and nonpoint sources, namely urban, agrarian and floods; (c) these nutrients travel from their sources via surface water and groundwater flow to the lagoon.

Assuming this scientific diagnosis, we performatively situated the exercise as a formal collaboration from all public administrations involved in the governance of the Mar Menor, and their commitment to the recommendations resulting from the deliberation. We then introduced role-playing as a way to simulate a random selection process for participants in the assembly. Random selection is a crucial element of CAs and mini-publics. It follows a sortition process statistically structured to represent the whole population affected by the mandate and to guarantee equal probability of participation (Flanigan et al., 2021). We designed 25 roles to cover a range of Mar Menor "inhabitants" "types", giving them a short context on age, gender, profession and opinion about the lagoon's situation. Considering that participants in the training were mostly members of environmental organizations, the roles focused on other profiles living in the Mar Menor area: fishermen, people working on the services and tourism sectors, students, retirees, housewives, small farmers, migrant workers in the agricultural sector, owners of agroexport companies, representatives of agricultural organizations, public servants at administrations and five other-than-human beings, including the lagoon itself. Participants without roles could participate as their own person. The purpose of giving roles to only half participants was to enable conversations and learnings happening at two different levels, concerning personal experiences and concerning the diversity of roles in the Mar Menor problem. Finally, we considered that experimenting with other-than-human roles could spark reflections relevant to the emerging governance structure for the lagoon's legal personhood.

For participants to embody their roles, we guided an immersive exercise. We asked them to move throughout spaces and find a spot that for them represented the place where this role or themselves occupy in relation to the Mar Menor. Then, we asked participants to internally visualize their answers to the following questions: Who and how are you in this place? How is the landscape around you? Is anybody with you? Where is the Mar Menor, close or far away? Are you affected by its crisis? Are you angry about what happens? Or are you rather indifferent?

Participants moved to one of five dialogue groups of about ten people that we preassigned, mixing half of participants role-playing and half not. Each group was tasked with a specific topic related to eutrophication: agriculture, cattle production, urbanisms, floods and biodiversity conservation in the lagoon. They were tasked with identifying required changes for a less eutrophic lagoon within their group topic, and to classify those changes according to the following typology: individual, relational, cultural and structural. After that, they had to identify policy measures and concrete actions. To support them in this task, some experts on the Mar Menor eutrophication attended their doubts and questions.

Deliberation lasted two hours followed by a plenary session to share and discuss proposed measures. Then, we asked participants to rank the order of priority for all measures in high, medium, low. By the end of the morning, participants were invited to write a short reflective report addressing the following questions:

- What did you feel (distinguishing role players and not role players)?
- Was there anything that made you feel uncomfortable or irritated? (could be an opinion, another person's attitude, the group dynamic…)?
- Was there anything that made you feel joy or excitement? (could be an opinion, another person's attitude, the group dynamic…)?
- Do you think your opinion about how to address the eutrophication problem has changed along the day? If affirmative, what and how changed?

Participants shared their personal experiences and learnings based on these reports during the afternoon reflexive session.

Recommendations from the Mar Menor simulated assembly

Participants discussed intensively on the changes and actions needed for a less eutrophic lagoon. They proposed up to nine overall changes, six of which were classified as structural, one as relational, one as cultural and one as individual. Four of the changes referred to compliance with and implementation of existing regulations. Two asked for new regulations on land planning.

Concerning concrete recommendations, three focused on the transformation of the agricultural and cattle production model "from one oriented to international exportation to another based on local markets, social justice and environmentally clean". This recommendation was the most highly rated by participants, and sparked debates, such as: Better salaries and fair conditions for agricultural workers, don't they increase costs for small producers? Shift in agricultural practices, but how exactly (reducing cropping area, shifting to organic farming)? How much agriculture or cattle production can be "sustainable" in the Mar Menor? Beyond these discussions, a CA with the agricultural sector was proposed as an opportunity to agree on a new legislation to enable such a transition.

The group on biodiversity protection recommended "more life in the Mar Menor" by eliminating all discharges from human activities, creating stricter regulations and controls, and fining and expropriating land for those who fail to comply. The group on floods was quite conflictive. They agreed to disagree and did not recommend any change. Finally, the group on urbanism recommended changes in education at different levels that could be funded by programs on property taxation and improving land-use plans.

In the plenary discussion, all persons playing more-than-human roles spoke up. They seemed relaxed and playful. Some in fact displayed movements or spoke about their feelings as plants and animals. The lagoon's role gathered a lot of attention and raised comments about who speaks on its behalf and how, who represents "its dignity", who is "the doctor caring for its health" or "giving it a human voice, isn't it a way of anthropocentring the lagoon?". There were also questions on how to bring the voices of children and of future generations.

Collective learning about CAs and conflicts

In this section, we provide a summary of the conversations held in the afternoon reflexive session. We describe four moments that speak to the goals of the session, that is, those that tell something about the social conflict around the Mar Menor and about the presence of more-than-human voices.

The *first conversation* was around Antonia's[1] experience within the flooding group. She is a leading figure in one environmental organization. She is passionate and energizing. She played the role of the head of the foundation lobby for the agricultural sector. She goes:

> the role has possessed me and I felt rejection from my colleagues, to the point that I asked to stop representing the role. Even when I stopped playing, they rejected anything I proposed. And I said to myself: what is this?

She continues:

> We noticed our discussions ended nowhere and we reflected on how difficult it is to come to consensus, and how we could change the role that I was embodying. It is difficult to move someone in such a position because what she receives is hostility, she won't change or open a fracture in her opinion.

Antonia's reflection speaks directly to what we expected to happen, those roles that are perceived as enemies by a majority of environmental defenders would face hostility and rejection. This power dynamics would probably reverse if the majority of participants in the assembly were farmers and still remain the same: marginalization of the minority. Another participant in her group shared "It was frustrating to contest her, she answered with arguments pretending they were scientific, because she had her experts". Yet, Antonia felt a victim as her colleagues confronted her: "I was a technician of the public administration and I felt I had to signal the flaws in her proposals". This person represented the voice of authority, that who knows how things work, which introduces another asymmetry between who knows and who doesn't, whose knowledge is valid and whose is nonsense.

The point here is not to arbitrate who is right or wrong. The point is that these dynamics are often invisible yet shape collective interactions. When the facilitator asked how she felt about what happened, Antonia replied: "Bad, really bad, to the point of quitting the role. And yet they kept disagreeing with me. That got me thinking, maybe we are not as aligned as I thought, maybe we do not share the same goal".

The *second conversation* was engendered around Teresa's experience playing the role of the lagoon in the biodiversity group: "I felt embraced by the other participants, they were happy to have me in the group. Nevertheless, I felt really anxious because they were not giving me any solution".

The facilitator asked: "You felt you had no agency by yourself?", "No, I was in their hands, I depend on them, and they have their own needs", replied Teresa. Then the facilitator opened the floor to the experience of the rest of other-than-human role players. The person playing the seahorse added: "I felt lonely, extremely lonely, and sad". Teresa, the lagoon, adds: "It was like, what you are saying is ok, but it is not enough".

Later in the session, Antonia came back to these comments to defend the importance of empathy with nature: "We as citizens have to represent not the farmers, nor the cattle producers, but the lagoon, we need to give it a voice, those of us who do not seek economic benefits, who can think out of economic benefits".

The facilitator asked her: "Maybe you obtain other types of benefits..?", another participant replied: "Some of us do not just want to enjoy the beach, we would be satisfied simply with seeing the Mar Menor better. We are not thinking about ourselves but in nature, what would be this kind of interest?"

This conversation speaks of relational values to nature (Chan et al., 2018) and how they are instrumental in the discussion of giving a voice to the lagoon. It also shows how the Mar Menor appears in participants' experience as a subject whose voice needs to be elevated against those who neglect it in defense of economic interests. Environmental defenders feel compelled to play that voice. Yet, the person playing the lagoon felt urgency and insufficient diligence in the proposals of those that pretend to defend her, opposed their needs to hers. It was a demanding voice indeed.

The *third conversation* was sparked by Ana's role as a woman, migrant from Ecuador, working in a lettuce packaging industry at Campo de Cartagena. She participated in the group on agriculture. She first shared that she had been working on human rights for migrants during the past twenty years, and therefore she's quite sensitive to this type of reality.

> I got really angry. I had a very clear idea of this role, I even chose a name, Clara. I am a woman that enjoys bathing in the Mar Menor with her children and thinks tourism is very important despite working in the agricultural sector. When I raised this opinion, I was interrupted and told this is a group for talking about agriculture, my opinion did not count. And when I said the discussion was very technical and I did not understand, nobody explained anything to me. And when I brought an argument for labor rights, I felt condescension towards my role, like, ok we include this, like it was my issue because I am a migrant, not because of a problem of power and economic class (…) So this is what happens with this voice, it's not present, it has not been present in the whole training on the Mar Menor. I missed it. The voice of those who sustain agriculture, cattle production, building… or who else sustains the world at this moment? We did not hear about them in this training. If I were truly Clara, I would leave and not come back.

Others in her group acknowledged what she described. The facilitator signaled she was embodying intersecting power inequalities in her role, "this is what sometimes happens with marginalization, people just leave". Ana's comment opened a long conversation on the invisibility of migrant experiences in the Mar Menor problem. "What do we do with the precarious, those who are damned, those who do not have anything, aren't they part of what we speak when we speak of nature?" She concluded. Someone echoed

> ...migrant population is perceived as an appendix of productive activities, they are not rooted here, and they are not perceived to have political rights because they can't vote, it's like they do not count. It's like they are not there in the Mar Menor problem, they are not actors.

In our view, this was a very powerful conversation that speaks to structural inequalities and the barriers for marginalized people to have a voice in public matters. Migrants are not an actor in current stakeholder analyses of the Mar Menor. They are often considered a vulnerable group, but invisible to dominant discussions around nutrient enrichment. Would they be selected for a CA? Those without a regular situation would not. Even if selected, would they be treated as equal participants?

The *fourth conversation* was around complaints from Jorge. He felt forced to play the role of a rancher, and marginalized during discussions, within the group on cattle raising. He was expecting to discuss solutions to the Mar Menor problem from his own personal perspective in this session. He was double disappointed by playing a role and by not arriving at a clear recommendation because there wasn't enough time.

Another participant reacted "sometimes dialogue is not fruitful, you think you're talking of the same question but they are perceived differently". The facilitator reflected back on the importance of managing expectations and on the challenges of dialogue spaces: "many people come to dialogue spaces to defend their position and their interests, they are told to listen, but, is that enough? The willingness to reach an agreement or to transform my own opinion is something different".

This conversation speaks of two insights. First, there is a tension between producing recommendations and dealing with relational dynamics like marginalization in a short time. CAs usually last several weekends, yet they have a heavy burden of knowledge transference and deliberation is oriented to producing the recommendations. Engaging with conflictive dynamics requires care and time which may be at odds with arriving at the desired product. Second, even providing space for controversial discussions, coming to an agreed recommendation might not be possible.

What can we learn for a CA in the Mar Menor?

Deliberative mini-publics, and CAs in particular, are called to regenerate the democratic space by bringing citizens back into public matters (Ganuza Fernández &

Mendiharat, 2020). They have proven a novel democratic format to overcome political paralysis in complex and polarized policy issues (Suiter et al., 2022). However, when applied to environmental problems with multiple dimensions, knowledges and actors involved in confronted positions, there is still need for experimentation and learning (Cherry et al., 2021). This chapter systematizes an epistemic experiment on a CA focused on how to shift the Mar Menor to a less eutrophic state. Within a flexible and playful participatory space, the exercise aimed to contribute to experiential "whole-person" learning (Craps & Brugnach, 2021) concerning two open discussions within the climate CAs community: first, the challenge for deliberation within conflictive environments (KNOCA, 2023a), and second, how to listen to other-than-human experiences (KNOCA, 2023b).

The experience of simulating a CA enriched participants' symbolic construction of the environmental conflict in the Mar Menor. This was possible because both the previous sessions of the training and the role-playing honed their view of the social-ecological crisis and of the multiplicity of actors entangled in it. Also because group work illuminated the intricate dynamics of cooperation and conflict that emerge among diverse actors when engaged in open dialogue (Bodin et al., 2020). Furthermore, involving individuals from environmental organizations in CA simulations and experimenting with different positionalities opens up the possibility of expanding their capacities for empathy and dialogue.

Concerning the recommendations proposed, it is noteworthy that most targeted compliance with the legislation and the transformation of the agricultural sector. This is unsurprising considering the nature of the eutrophication problem and the environmental positionality of participants. However, many open questions remain about what actually means to change the agricultural model and how a CA can help enact such a shift.

Groups dynamics during deliberation were diverse but overall show that a real CA in the Mar Menor would not be an easy task. Contestation dynamics will soon emerge as they did in our exercise. Extra time and care would be required for working on relational aspects and for going deep into controversial topics. Furthermore, reaching agreements and final recommendations is not guaranteed. Our experiences with conflict transformation processes, and plenty of literature, suggest that approaches based on respecting difference and dissensus rather than forcing consensus may be more effective in finding bridges among polarized positions (see, for instance, Capizzo, 2023; Gorostidi-García et al., 2023; for an example within the Mar Menor context see Cabello & Zuluaga-Guerra, 2023). Yet, if recommendations are not the output of a deliberative mini-publics, can it still be named a CA?

A significant insight from this exercise is the empathy developed for both other-than-human entities and marginalized human groups such as the migrant population in the Mar Menor. Recognizing the feelings of exclusion or disenfranchisement of migrant experiences highlights the challenges of including the most invisible in the Mar Menor and many other territories in CAs. As our conversations have bluntly shown, they are not considered an actor, an agent with an experience that needs to be heard and attended as equal.

On the other hand, the lagoon emerged as a central subject whose voice does count. This brings its own challenges as there are reasonable criticisms to speaking on behalf of those living experiences we cannot attain (Noorani & Brigstocke, 2018). Our insights speak to feelings of loneliness and impotence when trying to step into those shoes, but also of desires of elevating this subject as a citizen with rights. How then to open the space for the lagoon in its own agency is a relevant research question and a field for experimentation.

The reflexive analysis of the experienced process not only facilitated a deeper comprehension of conflict and marginalization dynamics but also promoted transformations toward environmental peace (Hardt & Scheffran, 2019). For example, we observed that making visible and attending to the axes of discrimination that are operating in this given conflict facilitated a more profound learning among participants than the sole production of recommendations.

In conclusion, we argue that the Mar Menor offers unique opportunities for advancing democratic responses to the complex social-ecological challenges it faces through critical and emancipatory dialogue (Mouffe, 1999) that includes both human and more-than-human experiences. We suggest future deliberative spaces within the Mar Menor, be it CAs or other formats, take conflict and marginalization seriously and dedicate time and resources to learn how to enter, hold and transform conflictive interactions among the diversity of ways of knowing and experiencing ecological degradation.

Acknowledgments

The authors are grateful to Pablo Rodriguez Ros for supporting the organization of the learning experience presented here. This paper has benefited from the support of the following institutions: Spanish Ministry of Science and Innovation through the program Ramón y Cajal (RYC2021- 031626-I), the project 'Crisis climática, salud mental y bienestar en el Antropoceno. Una aproximación desde la ontología histórica' (PID2021-124477OA-I00 MCIN/AEI/10.13039/501100011033 and 'ERDF A way of making Europe'), and the project 'Construyendo la sociedad sostenible. Movilización, participación y gestión de prácticas socio-ecológicas' (PID2021-126611NB-I00 MICIU/AEI /10.13039/501100011033 and FEDER, UE); the María de Maeztu program for accreditation of excellence 2023-2027 (CEX2021-001201-M); and the Basque Government through the BERC 2022-2025 program

Note

1 Names have been anonymized.

References

Bodin, Ö., Mancilla García, M., & Robins, G. (2020). Reconciling conflict and cooperation in environmental governance: A social network perspective. *Annual Review of Environment and Resources*, 45, 471–495. https://doi.org/10.1146/annurev-environ-011020-064352.

Boix-Fayos, C., Martínez-López, J., Albaladejo, J., & de Vente, J. (2023). Finding common grounds for conflict resolution through value analysis of stakeholders around the socio-ecological crisis of the Mar Menor coastal lagoon (Spain). Landscape and Urban Planning, 238, 104829. https://doi.org/10.1016/j.landurbplan.2023.104829

Boswell, J., Dean, R., & Smith, G. (2023). Integrating citizen deliberation into climate governance: Lessons on robust design from six climate assemblies. *Public Administration*, 101(1), 182–200. https://doi.org/10.1111/padm.12883.

Cabello, V., & Brugnach, M. (2023). Whose waters, whose nutrients? Knowledge, uncertainty, and controversy over eutrophication in the Mar Menor. *Ambio*, 52(6), 1112–1124. https://doi.org/10.1007/s13280-023-01846-z.

Cabello, V., & Zuluaga-Guerra, P. (without order) (2023). Lessons Learnt from a Process of Co-research with Social Actors of the Mar Menor. Available at: https://www.shareddialogues.org/wp-content/uploads/2023/08/Learnings_MarMenor.pdf.

Capizzo, L. (2023). Managing intractability: Wrestling with wicked problems and seeing beyond consensus in public relations. *Public Relations Review*, 49, 102263. https://doi.org/10.1016/j.pubrev.2022.102263.

Chan, K. M., Gould, R. K., & Pascual, U. (2018). Editorial overview: Relational values: what are they, and what's the fuss about? *Current Opinion in Environmental Sustainability*, 35, A1–A7. https://doi.org/10.1016/j.cosust.2018.11.003.

Cherry, C. E., Capstick, S., Demski, C., Mellier, C., Stone, L., & Verfuerth, C. (2021). *Citizens' Climate Assemblies: Understanding Public Deliberation for Climate Policy*. The Centre for Climate Change and Social Transformations. https://orca.cardiff.ac.uk/id/eprint/145771/.

Craps, M., & Brugnach, M. (2021). Experiential learning of local relational tasks for global sustainable development by using a behavioral simulation. *Frontiers in Sustainability*, 2. https://www.frontiersin.org/articles/10.3389/frsus.2021.694313.

Crosby, N. (2021). *Citizens' Assemblies on Global Climate Change in the US*. Center for Democratic Development.

Escobar, E., & Elstub, S. (2017). *Forms of Mini-Publics: An Introduction to Deliberative Innovations in Democratic Practice*. New Democracy Foundation. https://www.newdemocracy.com.au/2017/05/08/forms-of-mini-publics/.

Flanigan, B., Gölz, P., Gupta, A., Hennig, B., & Procaccia, A. D. (2021). Fair algorithms for selecting citizens' assemblies. *Nature*, 596(7873), Article 7873. https://doi.org/10.1038/s41586-021-03788-6.

Ganuza Fernández, E., & Mendiharat, A. (2020). *La democracia es posible: el sorteo cívico y la deliberación para rescatar el poder de la ciudadanía*. Consonni.

Guaita-García, N., Martínez-Fernández, J., Barrera-Causil, C. J., & Fitz, H. C. (2022). Stakeholder analysis and prioritization of management measures for a sustainable development in the social-ecological system of the Mar Menor (SE, Spain). *Environmental Development*, 42, 100701. https://doi.org/10.1016/j.envdev.2022.100701.

Gorostidi-García, M., Rodríguez-Berrio, A., & Aristegui-Fradua, I. (2023). Dissensus as part of dialogue in organizational change processes: A case study in an NGO. *IJAR – International Journal of Action Research*, 19(2), 125–141.

Hardt, J. N., & Scheffran, J. (2019). *Construcción de paz medioambiental y cambio climático: evaluación, análisis crítico y perspectivas*. Pax crítica: aportes teóricos a las perspectivas de paz posliberal (1st ed., pp. 389–422). Tecnos.

Karakas, L. D., & Mitra, D. (2020). Believers vs. deniers: Climate change and environmental policy polarization. *European Journal of Political Economy*, 65, 101948. https://doi.org/10.1016/j.ejpoleco.2020.101948.

Kolb, A. Y., & Kolb, D. A. (2005). Learning styles and learning spaces: Enhancing experiential learning in higher education. *Academy of Management Learning & Education*, 4(2), 193–212. https://doi.org/10.5465/amle.2005.17268566.

KNOCA - Knowledge Network on Climate Assemblies. (2023a). Workshop on Hosting Climate Assemblies in Challenging Environments. Available at: https://www.knoca.eu/events/workshop-on-challenging-enviroments

KNOCA - Knowledge Network on Climate Assemblies. (2023b). Workshop on Hearing Unheard Voices: Listening to Future Generations and Nonhuman Nature in Climate Assemblies. Available at: https://www.knoca.eu/events/workshop-on-hearing-unheard-voices.

Jara, O. (2012). Sistematización de experiencias, investigación y evaluación: aproximaciones desde tres ángulos. *Revista Internacional De Investigación En Educación Global Y Para El Desarrollo*, 1(1), 56–70.

Mindell, A. (2014). Sitting in the fire: Large group transformation using conflict and diversity. Deep Democracy Exchange.

Mejía, M. (2012). *Produciendo el saber de la práctica. Sistematización. Una forma de investigar las prácticas y de producción de saberes y conocimientos.* Viceministerio de educación alternativa y especial del Estado plurinacional de Bolivia.

Mouffe, C. (1999). *El retorno de lo político: comunidad, ciudadanía, pluralismo, democracia radical* (M. A. Galmarini Rodríguez Trans.; 1st ed.). Paidós.

Noorani, T., & Brigstocke, J. (2018). More-Than-Human Participatory Research. *University of Bristol/AHRC Connected Communities Programme*, 44.

Pedreño Cánovas, A. P., de Castro Pericacho, C., & García, S. (2022). Producir la naturaleza: agricultura intensiva, estándares de calidad y controversias ambientales en el Mar Menor. In *La producción de la calidad en el sector agroalimentario: un análisis sociológico*, ISBN 9788419071590, pp. 17–78.

Ruiz, J. M., Albentosa, M., Aldeguer, B., Álvarez-Rogel, J., Antón, J., Belando, M. D., Bernardeau, J., Campillo, J. A., et al. (2020). Informe de evolución y estado actual del Mar Menor en relación al proceso de eutrofización y sus causas. *Instituto Español de Oceanografía (IEO)*. Accessed Mar 2024: https://www.miteco.gob.es/es/prensa/informe-marmenorjulio2020_tcm30-510566.pdf.

Scheidel, A., Del Bene, D., Liu, J., Navas, G., Mingorría, S., Demaria, F., Avila, S., Roy, B., Ertör, I., Temper, L., & Martínez-Alier, J. (2020). Environmental conflicts and defenders: A global overview. *Global Environmental Change*, 63, 102104. https://doi.org/10.1016/j.gloenvcha.2020.102104.

Sipos, Y., Battisti, B., & Grimm, K. (2008). Achieving transformative sustainability learning: Engaging head, hands and heart. *International Journal of Sustainability in Higher Education*, 9(1), 68–86. https://doi.org/10.1108/14676370810842193.

Suiter, J., Farrell, D. M., Harris, C., & Murphy, P. (2022). Measuring epistemic deliberation on polarized issues: The case of abortion provision in Ireland. *Political Studies Review*, 20(4), 630–647. https://doi.org/10.1177/14789299211020909.

Willis, R., Curato, N., & Smith, G. (2022). Deliberative democracy and the climate crisis. *WIREs Climate Change*, 13(2), e759. https://doi.org/10.1002/wcc.759.

Zuluaga-Guerra, P., & Cabello, V. (2023). El Mar Menor. Un relato colectivo. In Boj & Díaz (eds.) *Reset: Mar Menor. Laboratorio de imaginarios para un paisaje en crisis* (pp. 41–45). ISBN 978-84-127165-1-1.

9 The rights of Nature and the case of the Mar Menor. Implementation of Law 19/2022 of 30 September and enforcement in the judicial sphere

Teresa Vicente Giménez
and Eduardo Salazar Ortuño

The rights of Nature and the case of the Mar Menor

We are currently living in a new geological era of the Earth, the Anthropocene, in which human beings have become the greatest geological force on our planet. It is now, when we are facing an ecological crisis, which is part of a general crisis that includes the economic, social and political crisis, that we must once again ask ourselves about the place of human beings in the world and the state of our world. From the field of philosophy, Bruno Latour addresses the question "Where are we?" and states that "we are inside Gaia", that is, we are part of Nature, we are one more species in the ecosystem, but we are the only species that has managed to transform the Earth it inhabits: "We must think that the environment is made by living beings and not, as was previously believed, that living beings occupy an environment to which they adapt themselves… However, they have transformed everything: the minerals, the mountains, the atmosphere. They have transformed the conditions of existence". Being inside Gaia, inside which humans are able to modify habitability considerably, means that the fundamental question is now "the habitability of the planet" (Latour, 2023, pp. 56–57).

The current change of the world, which leads humanity to no longer inhabit the same Earth as before, places us in a new revolution comparable to the Galilean revolution that has displaced us from the great cosmology of the Moderns, and which leads us to affirm the end of Modernity. As Latour points out, the typical version of the Moderns is no longer possible:

> The Moderns also had a cosmology that allowed them a worldwide, global expansion. To simplify: it was a very particular cosmology of division, of the distinction between a <world of objects> and a subject that is somehow distant from them.
>
> (Latour, 2023, p. 34)

In the new worldview, the distinction of the moderns has disappeared: the human subject is ecology and the centre of life has shifted from the human to the ecosystem, in whose interaction humanity finds itself. From the field of the philosophy

DOI: 10.4324/9781003489078-10

of law, a new model of ecological justice opens up that recognises the interaction between humanity and ecology and the dignity of the Earth, of the ecosystem of the biosphere, recognising it as a subject of law.

The second Copernican turn implied by this new vision of the Earth and the place of human beings on it opens up a new philosophical, scientific, political and legal revolution. The Spanish philosopher Antonio Campillo, in his book *Grecia y nosotros. La herencia griega en la era global*, places special emphasis on the passage from the closed world of classical Greece to the infinite universe of modern philosophy, science and politics, and from there to the new current vision (Anthropocene) that leads humanity back to the Earth and its place in it. As he explains:

> In chapter five I begin with the scientific revolution of the 16th and 17th centuries, which was described by Alexandre Koyré as the passage <from the closed world to the infinite universe>, that is, from the Aristotelian-Ptolemaic kosmos to the Universum of Descartes and Newton, passing through Copernicus, Bruno, Kepler and Galileo. But this path leads us back to Gaia…that is, to a new vision of the Earth and the place of life on it -including the life of human beings themselves-.
>
> (Campillo, 2023, p. 21)

In the 15th and 16th centuries, European society and culture in the Modern Age experienced a new intellectual attitude in the way of understanding man and his superiority over Nature: humanism. The anthropological dualism that underlies modern humanism, centred on the human being as separate, superior and dominating Nature, has been clearly questioned today by a new generation of thinkers such as Lovelock, Arendt and Latour, in which the human being goes from being the dominating subject of Nature, converted into an object at his service, to being part of the Earth he inhabits, recognising his interaction with Nature. In this sense, Arendt's distinction between the modern epoch and the modern world, which Antonio Campillo picks up on, is very interesting: The "alienation of the Earth" initiated by modern science from the 16th and 17th centuries onwards has ended up becoming in the 20th century the most extreme form of modern "alienation of the world" and can destroy our "love of the world", for Arendt "love of the world was the first victim of the triumphal world alienation of the Modern Age" and has not only devalued our loving relationship with Nature but has also given rise to techno-scientific innovations that may endanger the continuity of human life on Earth. For this reason, Arendt distinguishes between the "modern era", which began in 1492, and the "modern world" which began in 1945 with the nuclear bombs dropped on Hiroshima and Nagasaki, a world, that of the second half of the 20th century, where human beings have acquired the power to destroy all life on earth, or at least human life (Campillo, 2023, pp. 217 and 218).

In the Modern Age, the great transformation of humanism was accompanied by the beginning of the scientific revolution of the 16th and 17th centuries and the formation of the sovereign state and modern capitalism in the 17th and 18th centuries. The rule of law or Western democracy was born at the end of the 18th century with the triumph of the bourgeois liberal revolutions – the French Revolution and the American Revolution – and was consolidated and developed in the 19th and 20th centuries, when it reached (in the second half of the 20th century) its most advanced model: social democracy, that is, the effective recognition of social justice through the consecration of universal human rights in the international sphere and social rights in the constitutional sphere of the States. The rule of law formula is linked in all its development to the 19th century and its liberal origin, which is why it cannot hide the correspondence between the rule of law and the market economy of capitalism, both in the liberal democracy that triumphed in the 19th century, and later, after the Second World War, in social democracy, which tries to reconcile full employment and social benefits with the capitalist market (Vicente, 2006, p. 100).

The rights of Nature emerged in the 21st century from ecological justice, just as social rights emerged in the 20th century from social justice. This advance in basic rights, now for ecosystems, implies an advance in equality and in the distribution of rights, it also implies a limit to the unlimited exploitation of natural resources whose great acceleration has taken place in the last years of our "modern world". In the 21st century, we no longer recognise ourselves as modern, because we recognise ourselves as inhabitants of a planet to which we belong but which does not belong to us.

When the initiative for a law as revolutionary as the one that recognises the rights of the Mar Menor arises from a social movement in response to the imminent collapse of the lagoon, it is because the ecological crisis has awakened a new ecological awareness. When this ecological awareness, this feeling of being one with the ecosystem to which you belong, manifests itself in a communal embrace of the lagoon to express that I am the Mar Menor and the Mar Menor is me, there is, at the same time, a move towards an ecological ethic. We realise that the recovery and conservation of the Mar Menor affects us greatly because we are Nature, that our physical and mental health depends on the health of fish, crustaceans, animals, plants and ecosystems, and vice versa, that respecting the needs of Nature is our responsibility, and we must address it for ethical reasons and for the functioning of life. In the riverside squares, the Plaza del Espejo in Los Alcázares became the epicentre, where the social movement was concentrated and politicians, scientists and ecologists gathered to explain why the collapse of the Mar Menor had occurred, and to understand that a model of development, production and consumption that has treated the lagoon ecosystem as an object at the service of economic profit is responsible for the damage caused.

The social movement, from its new vantage point of ecological awareness and ecological ethics, was ready for a new model of integrative justice not only between humans (social justice) but also between humans and non-humans: ecological justice. This was the reason why citizens welcomed with hope, joy and strength the new proposal, which I put forward in the Plaza del Espejo, to turn the Mar Menor into a subject of law, a legal personality, in order to guarantee it its own charter of rights, which includes the right to exist, to protection and to recovery.

As Nancy Fraser warns, "in a capitalist society, capital itself becomes a Subject", and she points out as one of the characteristics of capitalism "the inherent but blind directionality of capital, the process by which it constitutes itself as the subject of history, displacing the human beings who have created it and turning them into its servants" (Fraser, 2020, p. 18). Indeed, commercial companies have been recognised as legal entities and subjects of law since the 19th century, before women, children or indigenous peoples; and in our Civil Code, corporations, associations and foundations are recognised as subjects of rights in addition to individuals. In the 20th century, those excluded from liberal democracy, which enshrined capital and its agents as subjects of law, led large social movements demanding their status as subjects of law and came to enter the legal sphere with the new model of social justice and the positivisation of social rights common to all human beings, but not a word about the interrelationship between human beings and Nature. It was not until the 21st century, with the ecological crisis, that a worldwide social movement emerged to fight for the rights of Nature, born out of its needs, and the first normative texts appeared in which the rights of Nature were recognised on all continents.

Certainly, when we speak of an ecological crisis we are referring to a multiple crisis, with different dimensions – economic, social and ecological – that are intimately intertwined in the deep grammar of capitalist society. Each strand of this crisis, inextricably intertwined, lends itself to a critique centred on sustainability because the unbridled commodification of Nature is unsustainable and bound to harm both society and the economy. Now more than ever the question of Nature is at the heart of the current crisis as a whole. As Nancy Fraser states: In the 21st century, the commodification of nature has gone far beyond what one might think, the privatisation of water, the bioengineering of sterile seeds and patents on DNA. "Far from merely trading in existing natural objects, these forms of commodification generate new ones; probing deep into nature, they alter its internal grammar as much as the assembly line altered the grammar of human labour" (Fraser, 2020, p. 46). It is therefore not surprising that in recent years, social movements whose struggles are related to Nature have exploded, such struggles involve movements for emancipation. Consequently, for Fraser, critical theory for the 21st century must connect the critique of commodification with the critique of domination. The social movement that has fought for the rights of the Mar Menor has been for the emancipation of the Mar Menor from the hands of its unrestrained exploitation, and with its triumph, we have given Nature its manumission.

Law 19/2022, which recognises the Mar Menor as a subject with its own rights, is a step forward on the road to a new pact between people and Nature. It is a major change in law, an ecocentric shift that places life at the centre and allows us to transform our hegemonic thinking and act within planetary limits, to understand and respond to the serious social and human ecological crisis we are facing. As Yayo Herrero reminds us, we are eco-dependent and interdependent: "Human beings live embodied in a body that must be fed and nourished, and this, in turn, is embedded in a natural environment". And being eco-dependent beings, when considering the insertion of the human species in Nature, we are fully immersed in the fact that we are a planet with physical limits:

> There are nine planetary limits in the biophysical processes that are fundamental to guarantee the continuity of the processes of nature. Interdependent

> on each other, they provide a framework within which humankind can operate with some safety. Exceeding them places us in an environment of uncertainty from which large-scale and rapid changes can occur, leading to other natural conditions that are less favourable for the human species.
>
> (Herrero, 2023, p. 52)

Our condition as eco-dependent beings, that human beings are Nature, is translated into Article 6 of Law 19/2022, "the real keystone of the innovative system of protection granted to the Mar Menor", as defined by the Magistrate-Judge of the Instruction Court N.4 of Cartagena in his Ruling of the twenty-third of October two thousand and twenty-three (DPA Diligencias Previas Proc. Abreviado 231/2022). Thus, Article 6 literally states that "any natural or legal person is entitled to defend the ecosystem of the Mar Menor...before the corresponding Court or Public Administration".

Law 19/2022 recognises the Mar Menor of its right to life, to protection and recovery, and with it, has given it manumission, has emancipated the Lagoon and, with it, has opened the path to Peace with Nature, at a time so necessary, because we live in war, with ourselves and with Nature. We are not going to win the war, neither against each other because the nuclear weapons held by the conflicting powers will destroy us all nor against Nature because we need her for our survival but she can live without us. War is absurd.

Having understood what we will suffer if we persist in a global situation in which shortcuts to resolve the crisis of our civilisation do not work, we must move towards a radical change of attitude to transform the world from a world at war to a world at peace. However, this change of attitude is not easy to bring about, it is easier to flee forward, and not only because of the power of the socio-economic model of production and consumption that dominates the world but because of, as Fernando Valladares explains, our drift towards self-destruction: "We have a water, climate, biodiversity and pollution crisis. But instead of stopping to change, we run away from the problem, running towards an uncertain future, without wanting to know how uncertain it is". The tendency towards self-destruction, the death drive as the tendency of all living things to return to an inert, inorganic state, opposed to the life drive and indissolubly linked to both, is explained by Valladares in psychic and psychological terms:

> The death drive pushes the human being to take pleasure in destruction in an unconscious way, as well as being a key psychic motor in humans that seeks to bring the psychic state of the person to a state without tension, a state of rest that allows counteracting the instability and excitement inherent in the life of any person.
>
> (Valladares, 2023, pp. 288–289)

The legal recognition of a Nature with rights has been a question discussed by legal doctrine since the 1970s, following the publication of an article by the University of California Law Professor Christopher D. Stone entitled "Should the trees have standing? Thoward Legal Rights for Natural Objects", in which he wondered about the possibility of trees having legal rights (Stone, 1972, pp. 450–501). Fifty years after Professor Storm's article, as Jurist and Professor of Philosophy of Law at the University of Murcia, I have had the opportunity to draft a Proposed Law to recognise the legal personality and rights of the Mar Menor lagoon and its basin, located in the Region of Murcia. Thanks to citizen participation, through the procedure of the Popular Legislative Initiative (PLI) recognised in Article 87.3 of the Spanish Constitution, Law No. 16019 (October 3, 2022) has been achieved, making the Mar Menor Lagoon and its basin the first ecosystem in Europe with its own rights.

In the field of positive law, the enshrinement of the rights of Nature appeared for the first time in Latin America with the Constitution of Ecuador in 2008, followed by Bolivia with Law No. 071 of December 21, 2010: the Law on the Rights of Mother Earth; by way of Jurisprudence, it is worth highlighting the Colombian Constitutional Court's Ruling T-622-2016, which recognises the legal personality and rights of the Atrato River. In Europe, the rights of Nature, which have long been a theoretical proposal, have been introduced into positive law with Law 19/2022, of 30 September, for the recognition of legal personality for the Mar Menor lagoon and its basin. Throughout the 21st century, there has been an unstoppable succession of cases on different continents of the recognition of rights to ecosystems of great ecological value and in danger, as reflected in the periodic reports of the Secretary General of the United Nations General Assembly in relation to the Harmony with Nature Programme (http://www.harmonywithnatureun.org/).

In Europe, the movement for the rights of Nature is very much alive, led by Spain with the case of the Mar Menor Lagoon and its basin as the first ecosystem with its own rights in our continent. In the different countries, studies, meetings, discussions, processes and advances are taking place, as described in the book *Rights of Nature in Europe*, which deals with the arrival of the rights of Nature in Europe (García Ruales et al., 2024). Last September, 14 European countries came to the Mar Menor, to celebrate the event of the Confluence of European Waters and to pay tribute to the Mediterranean coastal lagoon for the conquest of its rights by the citizens, most of these countries have been fighting for years for the rights of their endangered ecosystems: Portugal (Tagus), France (Rhòne, Loire and Senne), Denmark, Holland and Germany (North Sea), Germany (Spree), Norway (Akerselva), Sweden (Klarälven), Montenegro (Moraca, Tara and Malva), Italy (Venice lagoon), Serbia (Drina). We are now working together to advance the EU study on the rights of Nature of the European Economic and Social Committee (Carducci et al., 2020), entitled, Towards an EU Charter of the Fundamental Rights of Nature (https://www.eesc.europa.eu/en/our-work/publications-other-work/publications/towards-eu-charter-fundamental-rights-nature), and to advance the proposal for a European PLI to make this theoretical proposal a reality.

The great virtue of the legal personality of the Mar Menor and its basin has been its achievement through a PLI, a mechanism of direct citizen participation, which empowers people in the achievement of recognising the rights of Nature.

The process of collecting 500,000 signatures for the PLI, as the maximum instrument of participatory democracy enshrined in the Spanish Constitution, has been exemplary, as described in detail in the third part of the book *Justicia ecológica y derechos de la Naturaleza* (Vicente, 2023).

From the end of October 2020, when the signature forms sealed by the Central Electoral Board were collected, until the moment of the delivery of 639,824 signatures to the Central Electoral Census Office on October 27, 2021, the citizens' movement around the PLI was growing and was warmly received in the squares and streets of the Region of Murcia and beyond – without a strong organisation, without funding and with the sole presence of volunteers taking to the streets, in the midst of the COVID-19 pandemic. And all of this was achieved within the period established in the Organic Law 3/1984, of 26 March, without using up the extraordinary period of three months granted by the Bureau of the Congress of Deputies.

On April 5, 2022, the Plenary of the Congress of Deputies approved the consideration of the PLI by a majority greater than the reinforced majority of two-thirds, the maximum recognised by our legal system. On July 13, 2022, the Ecological Transition and Demographic Challenge Commission of the Congress approved, also by a reinforced majority of 2/3, the Proposition of Law for the recognition of the legal personality of the Mar Menor lagoon and its basin. On September 21, 2022, the Proposition of Law was approved in the Senate with the support of all the parliamentary groups except VOX, the only three votes against out of the 268 votes cast. The Proposition of Law on the rights of the Mar Menor leaves the Senate through the main door, reinforced by the amendments presented by the different political parties in both the Congress and the Senate, which at all times counted on the Promoting Committee for its negotiation.

Implementation of Law 19/2022 of 30 September and enforcement in the judicial sphere

Despite the legal recognition since 1990 of protection figures for this unique ecosystem as a protected natural space at international (RAMSAR, ZEPIM), European (Natura 2000 Network) and regional levels (partially Regional Park, Protected Landscape and globally as a Wildlife Protection Area), the degradation of the Mar Menor has been unstoppable. This failure of preventive environmental law has been due both to administrative inactivity at all levels – state, regional and local – in terms of management, control and inspection of environmental protection duties, and to the preponderance of regulations that have promoted polluting urban, agricultural and livestock developments in its surroundings, the repeal of specific legislation for the lagoon (Law 3/1987, of 23 April, on the Protection and Harmonisation of Uses of the Mar Menor) and the uselessness of regulations with an anthropocentric profile.

Reactive environmental law is trying to attribute responsibilities through the use of environmental crimes, administrative sanctions and environmental liability mechanisms, based on the "polluter pays" principle, but any definitive and exemplary decision is taking too long. On the one hand, the *Topillo* case – which seeks to prosecute companies and authorities responsible for the pollution caused by private desalination discharges in the "Campo de Cartagena" for environmental crime – is still awaiting trial before the Provincial Court and the second part of the case is dismembered in the Magistrates' Courts of Cartagena, Murcia and San Javier. The environmental liability proceedings against seven agricultural companies are not yet finalised, and the administrative sanctions which should be accompanied by restitution of crops illegally established in the Mar Menor basin are beginning to be adopted at an excessively slow pace in recent years.

Despite the fact that as a result of the mass mortality of flora and fauna, specific legislative instruments were put in place for the protection and recovery of the Mar Menor (Law 3/2020, of 27 July) and plans relating to the Natura 2000 Network (Integral Management Plan for the protected areas of the Mar Menor, Decree 259/2019, of 10 October), the public, which had already been organised in numerous specific groups in defence of the lagoon since 2016, took up the proposal put forward by the Legal Clinic of the Faculty of Law of the University of Murcia, to propose a PLI that would recognise the legal personality of the Mar Menor and its basin, and endow it with its own rights (Vicente Giménez and Salazar Ortuño, 2024, p. 90).

Law 19/2022, of 30 September, on the recognition of the legal personality of the Mar Menor and its basin, is a law consisting of only seven articles, but it contains a great transformative approach with respect to the previous relationship of citizens with the ecosystem of the Mar Menor and its basin, as revealed in the preamble explaining the context prior to the advent of the law. The brevity of the PLI and its clarity made it possible for hundreds of thousands of subscribers to scrutinise it and for it to be subject to minimal amendments agreed upon in the parliamentary phase.

In turn, the seven precepts that make up the regulation have different genesis, nature and possibilities of enforcement. In other words, and although this circumstance is not expressly stated in the Law, it is clear from its content whether the articles can be directly applied or whether they require regulatory development.

The first article is an explicit and unambiguous recognition of the legal personality of the ecosystem of the Mar Menor and its basin, the embodiment of the ecocentric approach that must initiate the transition from considering the ecosystem as an object of anthropic exploitation to convert it into a subject of rights. This recognition, with performative implications, does not require any legal development and is directly applicable. However, as an explanatory amendment and in order to guarantee legal certainty, the literal delimitation of the Mar Menor's basin, both on the surface and in relation to the aquifers included, was subsequently introduced in the second paragraph.

Article 2 lists and describes the rights of the Mar Menor and its basin, i.e. the right to exist and evolve naturally, the right to protection, conservation and the right to restoration. This catalogue of rights, through which they are duly clarified, does not require regulatory development either and is a directly applicable article.

A separate comment deserves the third article, dedicated to governance, which establishes a complex organisation that provides representation and guardianship of the ecosystem. In addition to the ecocentric turn of Law 19/2022 and the arrival of the rights of Nature at the European level, in order to make them operational and guarantee them, the institutes and mechanisms provided for in the law also empower citizens to participate directly in decision-making that affects the Mar Menor and its basin. Thus, the great legal milestone of the manumission of the Mar Menor from being considered an object of exploitation, a natural resource, to being recognised as a subject, as a legal person, does not stop there but is in turn accompanied by a firm commitment to participatory democracy as the axis of representation of the ecosystem.

This is consistent with two circumstances that have to do with the origin of the Law: on the one hand, the distrust of the public towards the management of the Mar Menor ecosystem by the public authorities and, on the other hand, the profoundly democratic origin of the proposal through a PLI, as an instrument of direct participation of citizens in public affairs. Furthermore, the PLI is rooted in the principles, rights and obligations of the Aarhus Convention, an international treaty ratified by Spain in 2005, which represents a commitment to environmental democracy (Salazar Ortuño, 2019).

The Guardianship of the Mar Menor also represents an unusual boost to participation in the defence of the Mar Menor and its basin, in that it goes beyond the bodies provided for in previous regulations and which have not yet been set up.

The governance bodies of Law 19/2022 are grouped in what is called the Guardianship of the Mar Menor and its basin, and there are three of them: the Committee of Representatives, the Monitoring Commission (Guardians and Guardians) and the Scientific Committee. This Guardianship is minimally described in article three of Law 19/2022 and requires, in order to detail its organisation and operation, a regulation that is about to be approved by the State Government and which was informed by the Council of State, as the last step, on January 25, 2024. The draft Royal Decree that would develop the regulation of Law 19/2022 and the organisational corpus of a private nature was submitted for public information in May 2023.

The rules of procedure will foreseeably confer functions on the Board of Trustees, in which decisions will be taken by a majority of the chairpersons of the three constituent bodies, with the chairpersons representing the majority of each of the bodies. The rules of procedure shall also lay down provisions on the financing of the Guardianship and common organisational and operational provisions for all the bodies.

According to Law 19/2022, the Committee of Representatives will be composed of thirteen members, three from the State Administration, three from the Regional Administration and seven representatives of the citizens, who will be selected from among the inhabitants of the riverside municipalities. In the first mandate, according to the legislative amendment, it will be the members of the Promotional Committee of the PLI who will represent the citizens. The Committee of Representatives will propose actions for the protection, conservation, maintenance and restoration of the lagoon; it will also monitor and control compliance with the rights of the lagoon and its basin on the basis of the contributions of the other two bodies (Monitoring Commission and Scientific Committee).

The Monitoring Commission, which includes the guardians, has a broader social spectrum and its composition would include, according to the draft regulation, eight representatives of the municipalities with territory in the area of the ecosystem with legal personality and eleven representatives of affected economic and social sectors such as fishing, agriculture, livestock farming, trade unions, environmental defence, business, neighbourhood, youth and gender equality associations. The main functions of the Monitoring Commission are the dissemination of the law and the monitoring of its compliance, in order to make proposals to the Committee of Representatives.

Finally, the Scientific Committee will be made up of independent experts specialised in the ecosystem of the Mar Menor and its basin, appointed by the research centres mentioned in Law 19/2022, for a four-year term, and its mission will be to provide technical advice to the other two bodies, giving content to the rights of the lagoon and its basin by identifying indicators on the ecological state of the ecosystem and appropriate restoration measures.

The draft Royal Decree, the approval of which will enable the bodies of the Mar Menor Guardianship to be set up, has still not been approved by the State Government despite the fact that, as mentioned above, the last step has been taken: the opinion of the Council of State. The delay in the approval of the regulation prevents the Guardianship from exercising its functions of representation and defence of the Mar Menor, in the face of a eutrophic state of the lagoon that does not cease.

The opinion of the Council of State, number 1276/2023, included valuable recommendations for the drafting and coherence of the regulation, and some salvageable criticism regarding the lack of definition of the nature of the legal entity of the ecosystem and, therefore, of the Guardianship. We say salvageable because we understand that given the lack of express mention in Law 19/2022 of the public status of the ecosystem of the Mar Menor and its basin, and the absence of recognition of a "mixed" personality, the private or ordinary status of the ecosystem must be deduced (Ayllón Díaz-González, 2023, p. 55).

The opinion of the Council of State has also criticised the ultra-activity of the Promoter Commission of the PLI, although it does not justify, beyond the legal opinion on the exhaustion of the role of the Promoter Commission, the negative influence of such a decision on the conformation of the Guardianship and the Committee of Representatives.

In our opinion, the only article of Law 19/2022 that needs to be developed in order to be applied in practice is the third article, dedicated to the Guardianship of the Mar Menor. The rest of the articles can be applied directly. Both those already mentioned, the first and second, and the fourth and fifth, the latter dedicated to the consequences of private conduct and the actions of the Public Administrations that do not respect the rights recognised for the Mar Mor and its basin. Neither does the sixth article, which establishes a legal standing by representation open to any person, nor the seventh article, added at the legislative stage and which contains a list of obligations of the Public Administrations in relation to the ecosystem of the Mar Menor and its basin, need further regulatory development.

It should be made clear that, precisely because only the third article requires regulatory development, this lack of development should not be used as an obstacle to prevent the application of the rest of the precepts of the Law, which have been fully in force since its entry into force on October 3, 2022.

Article 6 also deserves special comment as it is a procedural novelty in the Spanish legal system and also in comparative law. Again based on the principle of promoting broad access to justice derived from the Aarhus Convention (UNECE, 1998), the rule establishes the possibility for any natural or legal person to act before the Administration or the Courts "on behalf of the ecosystem". In other words, any person may bring administrative or judicial actions in defence of the rights of the Mar Menor and its basin. He or she will be able to bring actions before any of the jurisdictions: civil, criminal and administrative, representing the interests of the Mar Menor. This provision implies something radically different from a public or popular action, as it is not an action in defence of legality or general interests. It is an action in which the party is the Mar Menor lagoon and its basin (Peñalver i Cabré and Salazar Ortuño, 2023, p. 345). In addition to this change in the procedural approach, Article 6 provides for a series of mechanisms to reduce or eliminate the financial barriers that are normally associated with access to justice in environmental matters, such as the costs of the process or bonds for the adoption of precautionary measures.

Granting legal personality to the Mar Menor and its basin means giving it "legal capacity" and "capacity to act", as another entity within the legal system. This new legal institution means not only granting rights to the Mar Menor and its basin, i.e. "legal capacity", but also the possibility of exercising them, which is called "capacity to act". These rights or faculties must not only be respected by other natural and legal persons, in a new relationship of equality with classic rights linked to development such as private property or freedom of enterprise, but also become a legal title for any citizen to promote their respect and safeguard (Article 6).

Article 6 responds precisely to the phrase commonly used by the citizens who have participated in the PLI: "I am the Mar Menor".

As Nature cannot claim the danger (*Achtung*), because this can only be claimed by people, in the case that the Mar Menor may suffer danger or may be affected by damage, it is the person who has the legal duty of consideration and respect for the ecosystem (*Be-achtung*). In this sense, Article 6 of the Law on the rights of the Mar Menor legitimises any natural or legal person to defend the ecosystem of the Sea, this is, of course, a "private action on behalf of the Mar Menor", a new and broader legal figure than the traditional "popular action" included in our Criminal Procedure Act for more than a hundred years.

In our opinion, Article 6 is the most far-reaching, together with Article 1 where the legal personality of the Mar Menor and its basin is declared and Article 2 where the rights of the Mar Menor and its basin are recognised, its right to exist as an ecosystem and to evolve naturally and the rights to protection, conservation, maintenance and, if necessary, restoration. And this is how it has been understood by the Judge of the Magistrate's Court N.4 of Cartagena, the first judge to call upon the Mar Menor as a subject of law to prosecute in a case where it may be endangered by a crime against the environment, and he does so by issuing the historic order of 4 September two thousand and twenty-three (DPA 232/2022), in which the Mar Menor, as an entity with legal personality, is offered actions through the call to certain public bodies and private organisations of special relevance in the care of the Mar Menor. This relevance is justified because the bodies and organisations called by the Judge have been chosen by the legislator, the law itself in Article 3 includes them along with others, to hold the guardianship of the Mar Menor, trying with such appeals to limit the very broad offer of shares in article 6 that allows any person to be called, as the Judge himself clarifies in the Ruling 23/10/2023 issued in Cartagena.

However, the Provincial Court of Murcia, Cartagena Section, does not understand the letter of Article 6 as a private action on behalf of the Mar Menor but as a popular action, and issued Order 52/2024 against the Ruling of 04/09/23 of the Examining Court no. 4 of Cartagena, upholding the appeal lodged by the defendant company. This Resolution states that:

> It must be considered, therefore, that the action referred to in article 6 of Law 19/2022, is nothing more than a repetition or reminder of the provisions of article 101 of the L.E.Cr, or, in other words, the exercise of the popular action by any citizen or entity.

The Resolution leaves without effect the appeals made by the examining magistrate. With this action by the Court, people who have been called to exercise private actions in the name and on behalf of the Mar Menor see in danger the extent of the right to judicial defence of the Lagoon and its basin recognised in Law 19/2022 and, after presenting before the Court an extraordinary appeal for cassation and nullity of proceedings, both denied, the Mar Menor has appealed in amparo before the Constitutional Court claiming its right to effective judicial protection.

References

Campillo, A. (2023). *Grecia y Nosotros. La herencia griega en la era global*. Madrid: Abada Editores.

Carducci, M., Bagni, S., Lorubbio, V., Musarò, E., Montini, M., Barreca, A., Di Francesco Maesa, C., Ito, M., Spinks, L., & Powlesland, P. (2020). Towards an EU charter of the fundamental rights of nature. Brussels: European Economic and Social Committee.

Díaz-González, J.M. Ayllón (2023). Sobre derechos de la naturaleza y otras prosopopeyas jurídicas: a propósito de una persona llamada Mar Menor. *Actualidad Jurídica Ambiental, número* (138).

Fraser, N. (2020). *Los talleres ocultos del capital. Un mapa para la izquierda*. Madrid: Traficantes de Sueños.

García Ruales, J., Hovden, K., Kopnina, H., Robertson, C. D., & Schoukens, H. (Eds.). (2024). Rights of nature in Europe: Encounters and visions. London: Routledge.

Herrero, Y. (2023). *Toma de Tierra*. Bilbao: Caniche Editorial.

Latour, B. (2023). *Habitar la Tierra*. Barcelona: Arcadia.

Peñalver i Cabré, A. y Salazar Ortuño, E. (2023). Avances y retrocesos en acceso a la justicia en medio ambiente en la Unión Europea y en el Estado español. In *Observatorio de Políticas Ambientales 2023*. Agencia Estatal Boletín Oficial del Estado.

Salazar Ortuño, E. (2019). *El acceso a la justicia a partir del Convenio de Aarhus*. Cizur Menor: Aranzadi.

Stone, C. D. (1972). Should trees have standing?—Towards legal rights for natural objects. Southern California Law Review, 45, 450–501.

UNECE. (1998). *Convention on access to information, public participation in decision-making and access to justice in environmental matters* (Aarhus Convention). United Nations.

Valladares, F. (2023). *La Recivilización. Desafíos, zancadillas y motivaciones para arreglar el mundo*. Barcelona: Editorial Planeta.

Vicente Giménez, T. (2006). *La exigibilidad de los derechos sociales*. Valencia: Tirant lo blanch.

Vicente Giménez, T. (2023). *Justicia ecológica y derechos de la Naturaleza*. Valencia: Tirant lo blanch.

Vicente Giménez, T., & Salazar Ortuño, E. (2024). An ecological citizenship's triumph: From the popular legislative initiative to the rights granted for the Mar Menor. In J. García Ruales, K. Hovden, H. Kopnina, C. D. Robertson, & H. Schoukens (Eds.), Rights of nature in Europe: Encounters and visions (pp. 61–78). London: Routledge.

10 The Mar Menor (Spain) and incipient European regulations

From ecological restoration to agricultural flexibility

María Giménez Casalduero

Introduction

The Mar Menor is one of the most important coastal lagoons in Europe and the largest in the Iberian Peninsula. Its unique environmental values have earned it international, European, national and regional legal recognition by means of various forms of protection.[1] To date, these tools have not been able to prevent and control the numerous polluting activities carried out over the years in the territory, among other issues, due to the lack of application of the current legal system and the violation of the principles of prevention, precaution and non-regression of environmental regulations (Gutierrez Llamas & Giménez Casalduero, 2009; Giménez Casalduero et al, 2021).

Recently, after lengthy negotiations, Regulation (EU) 2024/1991 of the Parliament and of the Council of 24 June 2024 on nature restoration and amending Regulation (EU) 2022/869 was adopted, which is directly applicable to the Member States. This regulation, so-called Nature Restoration Act, includes an agreement to restore at least 20% of the EU's land and marine areas by 2030 and all ecosystems in need of restoration by 2050.[2] The main objective is to contribute to the continuous, sustained and long-term restoration of biodiversity-rich and resilient nature in all EU land and marine areas through the restoration of degraded ecosystems, be they wetlands, forests, marine ecosystems, agroecosystems, rivers and lakes and alluvial habitats, etc. At the same time, member state representatives in the Special Committee on Agriculture endorsed on 26 March 2024 a specific review of certain aspects of the Common Agricultural Policy (CAP) proposed by the European Commission, in response to the concerns expressed by farmers at various protests across Europe.[3] Among other issues, this review aims to address the problems identified in the implementation of the CAP's strategic plans and seeks to simplify, reduce the administrative burden and offer greater flexibility to comply with certain environmental conditionalities.

In relation to the Mar Menor, the usefulness of the Nature Restoration Act is manifested in the implementation of an essential measure: obliging Spain to establish legally binding tools to restore ecosystems and maintain them in good condition. For its part, the revision of certain elements of the CAP Strategic Plans Regulation and on financing, management and monitoring (the so-called "Horizontal

DOI: 10.4324/9781003489078-11

Regulation") seeks to balance the need to maintain a high level of ambition on environment and climate and to ensure that farmers' concerns are addressed. This ambitious scenario does not enjoy unanimous support among the different actors involved, both in the agricultural and environmental sectors. In recent years, the narratives and public dispute over the causes and solutions to the ecological crisis of the Mar Menor lagoon have become increasingly polarised (Cabello & Brugnach, 2023). Focusing on the issue at hand, the socio-ecological decline of the Mar Menor, the causes of its deterioration stem largely from the abandonment of low-impact extensive agriculture, the intensification of high-impact agricultural management practices, the modification of hydrological regimes, urbanisation and pollution (Esteve Selma et al, 2016). In this context, the Spanish State will have to draw up a national restoration plan, which will be regularly and strictly monitored by the Commission. The effects of the reduction of Good Agricultural and Environmental Condition (GAEC) standards proposed by the European Commission will also need to be monitored.

In this sense, the future scenario for the Mar Menor is uncertain and convulsive (Comité de asesoramiento científico del Mar Menor, 2017) The only reality is that to date neither the central government nor the Autonomous Community of the Region of Murcia, each in its own sphere of responsibility, has managed to prevent the ecocide of the Mar Menor. This work aims to clarify the fluctuating regulatory framework that European legislation imposes on Spain and, therefore, on the management of such unique ecosystems as the Mar Menor catchment basin and the economic activities that depend on it.

Legal elements for nature restoration in the EU

After a long process of debate under the EU's ordinary legislative procedure, the European Parliament voted in favour of the Nature Restoration Act in February 2024. However, its final adoption took place months later by all Member States at the EU Council. The main objective of this regulation is to ensure the restoration of ecosystems throughout the EU, which requires the establishment of EU-wide rules.[4] As set out in the Nature Restoration Act, the December 2022 Conference of the Parties to the Convention on Biological Diversity set a target of ensuring that by 2030 at least 30% of areas of degraded terrestrial, inland water and coastal and marine ecosystems are effectively restored, with the aim of enhancing biodiversity and ecosystem functions and services and ecological integrity and connectivity. This should preserve ecosystem functions and services, such as air, water and climate regulation, soil health, pollination and disease risk reduction, as well as protection against natural hazards and disasters, through nature-based solutions and/or ecosystem-based approaches for the benefit of all people and nature.[5]

In this context, the EU took up the baton and established an ambitious plan to restore nature based, among other commitments, on presenting a legally binding proposal to restore degraded ecosystems, especially those with the greatest potential for carbon capture and storage, as well as to prevent natural disasters and reduce their impact when they occur.[6] In this way, based on the European Green

Pact,[7] the EU aims to promote economic and social transformation, high-quality job creation and sustainable growth. However, without the implementation of an overall objective based on the restoration of biodiversity-rich ecosystems, such as wetlands, freshwater, forest, agricultural, sparsely vegetated, marine, coastal and urban ecosystems, it will be difficult to achieve the expected results. Clearly, essential ecosystem-based services will help to generate a wide range of socio-economic benefits, depending on economic, social, cultural, regional and local characteristics.[8]

In conclusion, by 2030, member states should prioritise the implementation of restoration measures on areas of habitat types that are not in good condition, and which are in Natura 2000 sites.[9] However, this does not preclude compliance with the obligation to establish restoration measures outside such sites.

Main obligations of member states in the field of nature restoration

According to article 3.3 of the Nature Restoration Act, restoration is defined as

> (…) the process of actively or passively contributing to the recovery of an ecosystem to improve its structure and functions, with the aim of conserving or increasing the biodiversity and resilience of the ecosystem, by enhancing an area of a habitat type to a good condition, re-establishing a favourable reference area and improving the habitat of a species to a sufficient quality and quantity (…).

On this basis and in line with the respect of the participatory principle, when drawing up and implementing their "national restoration plans", member states should involve local and regional authorities, landowners and land users and their associations, civil society organisations, the business community, the research and education communities, farmers and ranchers, fishermen, foresters, investors and other relevant stakeholders, as well as the general public, at all stages of the preparation, review and implementation of national restoration plans, and to encourage dialogue and the dissemination of scientific information on biodiversity and restoration benefits.[10]

Among the restoration objectives and obligations, the article 4 addresses terrestrial, coastal and freshwater ecosystems. In this sense, each State must establish the necessary restoration measures to improve, until they are in good condition, the areas of the habitat types listed in annex I. The text of the regulation is referring to all terrestrial, coastal and freshwater habitat types listed in annex I of Directive 92/43/EEC, as well as six groups of such habitat types, namely: wetlands (coastal and inland); grassland and other pastoral habitats; river, lake, alluvial and riparian habitats; woodland; steppe, heathland and scrub habitats; and rocky and dune habitats. Coastal lagoons obviously form part of this list of priority restoration sites.

When drawing up national restoration plans, article 14.14 requires member states to take into account, inter alia, conservation measures established for Natura 2000 sites under Directive 92/43/EEC; measures to achieve good quantitative, ecological and chemical status of water bodies included in the programmes of measures and river basin management plans established under Directive 2000/60/EC (Water Framework Directive) and flood risk management plans established under Directive 2007/60/EC of the European Parliament and of the Council ; where appropriate, marine strategies established with a view to achieving good environmental status in all marine regions of the Union drawn up in accordance with Directive 2008/56/EC[11]; national biodiversity strategies and action plans drawn up in accordance with article 6 of the Convention on Biological Diversity; where appropriate, conservation and management measures adopted under the Common Fisheries Policy (CFP); and, CAP strategic plans drawn up in accordance with Regulation (EU) 2021/2115.

Regarding the task at hand, the restoration of the Mar Menor catchment basin, national restoration plans will play a predominant role in the coming years. Among other aspects, it will be of utmost importance that adequate public and private investments in restoration are made and that member states, particularly Spain, integrate expenditure on biodiversity objectives into their national budgets, also about the opportunity and transition costs of implementing national restoration plans.[12] In the case of the Mar Menor, the obligation to restore should not focus exclusively on the coastal lagoon—a habitat listed in annex I of Directive 92/43/EEC and annex I of the Nature Restoration Act—but on the whole area of influence represented by the catchment area, which is plagued by economic uses and activities that directly affect the wetland. This is because the main cause of the current state of degradation of the Mar Menor is not in the lagoon but in its catchment basin, since the intensive irrigation of the Campo de Cartagena is the main cause of the eutrophic crisis (excess of nutrients, i.e. nitrogen and phosphorous, which causes explosive growth of phytoplankton) (Martínez et al, 2017; Martínez & Esteve, 2019).

Another important issue that the regulation introduces into the process of drawing up national restoration plans is that it allows member states to use the restoration measures listed in annex VII, depending on national and local specificities and the latest scientific data. These measures include the following: improving hydrological conditions by increasing the quantity, quality and dynamics of surface water and the water table for natural and semi-natural ecosystems; removing unwanted encroachment of scrub or non-native plantations in grasslands, wetlands, forests and sparsely vegetated land; restoring natural sedimentation processes; establishing riparian protection, such as forests, buffer strips, meadows or riparian grasslands; where necessary due to climate change, support migration of provenances and species; introduce landscape elements of high diversity on intensively used farmland and grassland, such as buffer strips, field boundaries with native flowers, hedges, trees, small woods, terrace walls, ponds, habitat corridors and stone walkways; increase the area of farmland subject to intensive cropping and grazing patterns; increase the area of farmland subject to intensive cropping and grazing patterns, such as hedges, hedgerows, trees, small woods, terrace walls, ponds, habitat corridors and stone walkways, etc. increase the agricultural area under agro-ecological management models, such as organic farming or agroforestry, multiple cropping and crop rotation, integrated pest and nutrient management; stop or reduce the use of chemical pesticides, as well as chemical fertilisers and animal manure; improve connectivity between habitats to allow species populations to develop and to allow sufficient individual or genetic exchange, species migration and adaptation to climate change; allowing ecosystems to develop their own natural dynamics, e.g. by abandoning exploitation and promoting habitats and natural areas; removing and controlling invasive alien species and preventing or minimising the introduction of such species; and, converting brownfields, former industrial sites and quarries into natural areas.

But undoubtedly one of the most interesting aspects of the regulation is the obligation to improve connectivity between the habitat types listed in annex I—terrestrial, coastal and freshwater ecosystems—as well as the ecological requirements of the species present in these habitat types. For freshwater ecosystems, such as the one identified in the catchment area of the Mar Menor, article 9 of the Nature Restoration Act obliges to ensure the maintenance of the natural connectivity of rivers and the natural functions of restored floodplains by removing artificial barriers and implementing the necessary measures to improve their natural functions.

The importance of the restoration of agricultural ecosystems for the Mar Menor

In the case of the restoration of agricultural ecosystems, article 11 of the Nature Restoration Act obliges member states to implement the necessary restoration measures to enhance their biodiversity, considering climate change, the social and economic needs of rural areas and the need to ensure sustainable agricultural production in the Union. This obligation is interpreted as a very ambitious proposal, especially where the agricultural area has grown exponentially in recent decades because of the transformation from rain-fed to irrigated agriculture. For example, in the Campo de Cartagena area, several remote-sensing studies have shown that between 1988 and 2009 alone, irrigation in the basin increased from around 25,150 ha to more than 55,000 ha, more than doubling (Carreño, 2015). The conversion to irrigation has been reactivated in recent years, with an estimated 15,000–20,000 ha of irrigated land outside the official figures. Given that each member state should identify and map agricultural and forestry areas in need of restoration, particularly those which, due to intensification or other management-related factors, need greater connectivity and landscape diversity, the catchment basin of the Mar Menor may benefit greatly.

Furthermore, as the preamble of the regulation indicates, the restoration of agricultural ecosystems has positive effects on food productivity.[13] For example, agricultural ecosystems rich in biodiversity will increase the resilience of agriculture to climate change and environmental risks, while ensuring food security and the creation of new jobs in rural areas, particularly those related to organic farming, as well as rural tourism and recreation.[14] It is therefore a priority need to enhance the biodiversity of agricultural land in the Union through a range of practices conducive to or compatible with such enhancement, including through extensive farming. As the text of the Nature Restoration Act makes explicit, "(…) such practices are not intended to stop agricultural land use, but to adapt agricultural land use to benefit the long-term functioning and productivity of agricultural ecosystems". However, to realise the long-term benefits of restoration, appropriate funding schemes must be available to enable landowners, farmers and other land managers to engage in such practices on a voluntary basis.[15] Indeed, member states may promote public or private support for the benefit of stakeholders implementing the restoration measures referred to in articles 4–12, including landowners and land managers, farmers, foresters and fishermen. What is interesting about the obligation to restore agricultural ecosystems is that they do not necessarily fall within the scope of Directive 92/43/EEC.

The Nature Restoration Act therefore introduces a general obligation to improve the biodiversity of such ecosystems based on a selection of indicators to be applied by member states. These indicators are based on the grassland butterfly index, organic carbon stocks in mineral soils on arable land or the proportion of agricultural land with high diversity landscape elements.[16] Regarding the indicator on "highly diverse landscape elements on agricultural land", its consideration for degraded areas related to wetlands and their surrounding area (e.g. the Mar Menor) is of particular interest. We refer to the case of buffer strips, rotational or non-rotational fallows, hedges, isolated or grouped trees, rows of trees, field boundaries, plots, ditches, streams, small wetlands, terraces, cairns, stone fences, small ponds and other cultural elements. They all provide space for wild plants and animals, including pollinators, prevent soil erosion and depletion, filter air and water, support climate change mitigation and adaptation, and agricultural productivity of pollination-dependent crops.[17]

In the Region of Murcia, most of these measures were introduced by Law 1/2018, of 7 February, on urgent measures to guarantee environmental sustainability in the Mar Menor, which introduced an extensive catalogue of good agricultural practices in its annex V. Undoubtedly, this regional law showed a high degree of courage in regulating its mandatory application throughout the Mar Menor basin. Unfortunately, the binding nature of these measures was abolished by Law 3/2020, of 27 July, on the recovery and protection of the Mar Menor. In fact, the current text in force contains an ambiguous formulation that opens the door to exceptionality and technical discretionality regarding the decision to consider certain measures classified as good agricultural practices in the basin as appropriate or not (Giménez Casalduero, 2022). In addition, today, around a dozen plans, programmes and strategies for environmental planning and management under Law 3/2020 are still pending approval and/or implementation. One of the documents still in the environmental assessment phase is the "Land Management Plan for the Mar Menor Catchment Basin", which regulates land use in ten municipalities in areas such as housing, agriculture and livestock.[18]

Agricultural flexibilisation and lowering of ecological restoration standards

One of the characteristic features of the CAP is the aim to support and strengthen the protection of the environment, including biodiversity, as its specific objectives include the need to contribute to halting and reversing biodiversity loss, enhancing ecosystem-based services and conserving habitats and landscapes.[19] Accordingly, within the scope of national restoration plans, member states should identify existing agricultural and forestry practices, including interventions under the CAP, that contribute to the objectives of the Nature Restoration Act.

With the adoption of Regulation (EU) 2021/2115 of the European Parliament and of the Council,[20] the new CAP cross-compliance standard no. 8 on GAEC of Land (GAEC 8) was introduced, as set out in annex III. This measure required beneficiaries of area payments to ensure that at least 4% of arable land, at farm level, was devoted to non-productive areas and features, such as set-aside land, and that existing landscape features were maintained. The 4% share attributable to compliance with GAEC 8 could be reduced to 3% if certain preconditions were met. The aim of this obligation was to help to achieve a positive trend in terms of highly diverse landscape elements on agricultural land. In addition, under the CAP, member states could establish ecological schemes for farmers' agricultural practices in agricultural areas, which could include the maintenance and creation of landscape features or non-productive areas. Similarly, member states were also required to include agri-environmental and climate commitments in CAP strategic plans, such as improved management of landscape features beyond GAEC or greening schemes.

However, due to the recent numerous protests by farm workers in many EU countries, farmers and livestock farmers have been demanding changes to the CAP for 2023–2027. Among other issues, their demands are aimed at denouncing excessive bureaucracy, the excess environmental restrictions imposed, the unacceptable weight, in their opinion, of CAP conditionality and the loss of competitiveness of the sector, compared to products from third countries that are not subject to EU restrictions. As a result of all these pressures, the EU and its member states are in the process of clarifying some of these aspects. Consequently, Commission Implementing Regulation 2024/587 provides for a derogation from Regulation (EU) 2021/2115 of the European Parliament and of the Council, as regards the application of the standard for GAEC condition of land (GAEC 8), and the rules applicable to the modification of CAP strategic plans. Specifically, article 12 of Regulation (EU) 2021/2115 required States to include in their CAP strategic plans (known in Spain as PEPAC) a cross-compliance system, according to which farmers and other beneficiaries receiving direct payments or annual payments would be subject to an administrative penalty if they did not comply with the statutory management requirements, in relation to climate and the environment—including water, soil and ecosystem biodiversity—; public health and plant health; and animal welfare. In the Spanish case, the regulatory basis on which these penalties are based is Royal Decree 147/2023 of 28 February, which establishes the rules for the application of penalties in the interventions contemplated in the CAP Strategic Plan and would basically consist of a reduction or withdrawal of the economic aid provided to the non-compliant farmer based on the committed eco-scheme.

As a result of the above discussions, the recent Commission Implementing Regulation 2024/587 clarifies the content of GAEC 8 on the minimum percentage of agricultural land devoted to non-productive areas or features. This standardises the criterion and allows farmers to comply with this basic condition by devoting a minimum percentage of their farmland of at least 4% to non-productive areas or features, including fallow land, and/or nitrogen-fixing crops and/or catch crops. In this way, as the Regulation itself recognises, "there will be fewer restrictions for farmers on how they use their arable land, and income losses will be reduced without forgoing the full environmental benefits". This concession by the European Commission has not been to the liking of all the actors involved in agriculture and livestock farming in the EU, and specifically in Spain. The coalition "PorOtraPAC"[21] has described the measures as "an unprecedented step backwards" that would make the CAP less environmentally ambitious than in 2014. According to the coalition, "It is irresponsible and will make it difficult for the agricultural sector to adapt to a very worrying climate scenario, the advance of desertification, the threat of drought and the worsening of nitrate pollution", the coalition said.[22]

Indeed, the response against the agricultural flexibilities has provoked the sending of a communication to the President of the European Commission by numerous European coalitions[23] and civil society organisations, urging her to abandon plans to dismantle the green architecture of the CAP and to present a systemic approach to respond to the diversity of farmers' protests and problems, encompassing the socio-economic, environmental and governance shortcomings of European agri-food and rural systems. In conclusion, these actors call on the Commission to adopt a coherent and evidence-based approach, in line with the EU's commitments on biodiversity and climate change, as well as with the targets set out in the EU Biodiversity Strategy 2030 and the Farm to Fork Strategy derived from the European Green Pact.[24]

Finally, on March 26, 2024, the member states' representatives in the Special Committee on Agriculture endorsed the changes to the standards on GAEC proposed by the European Commission.[25] Spain's response has also been to support these recommendations through the "Scoping Document on Amendments to the PEPAC",[26] with the approval of the European Commission. This has been welcomed and supported by many sectors of the agricultural sector, which consider that these flexibility measures solve an important part of the problems afflicting small- and medium-sized farmers. For example, the document exempts all farms of less than 10 ha from controls and sanctions for environmental conditionality, although this decision is not without risk. Linking this issue to the recovery of degraded ecosystems such as the Mar Menor, it is possible that this backtracking on environmental requirements will only benefit those who oppose changing the model of over-intensive agriculture. It is precisely this action that the scientific consensus points to as central to the recovery of the Mar Menor and a necessary condition for the success of the Nature Restoration Act in the area (Ruiz Fernández et al., 2019).

Conclusions

Nature restoration is a process of helping the recovery of an ecosystem that has been degraded, damaged or destroyed. According to the Nature Restoration Act, this process can consist of re-wetting wetlands and peatlands, improving degraded soils and agricultural land by adding natural elements such as hedges and trees, recreating lost natural forests, removing invasive species, planting native vegetation, removing obsolete river barriers, increasing tree cover in cities, etc. Despite conflicting positions on its application, nature restoration will be beneficial, not only in enhancing biodiversity and carbon sequestration, but also in increasing human health and well-being. In turn, nature restoration increases flood protection and water retention, as well as preventing forest fires. All this will make us more resilient to the effects of climate change. In fact, opponents of its implementation offer no solution to the accelerating climate and biodiversity crisis.

The recent measures adopted for the relaxation of agri-environmental practices related to the CAP are a concession to the agri-food sector that does not necessarily have to be negative. However, the lowering of environmental cross-compliance rules should not be negotiated at the demand of the market or the agribusiness sector. Healthy ecosystems are the basis for key ecosystem services, such as pollination, on which our food systems depend. Measures to enhance biodiversity on farmland should contribute positively to agricultural production through a range of benefits such as stable water retention, improved soil quality and healthy pollinator populations. This can be achieved by providing space for nature in agricultural landscapes in the form of uncultivated areas such as hedgerows, flower strips, wetlands or other habitats that are left (or managed) exclusively for wildlife.

In relation to the ecological crisis of the Mar Menor, the implementation of the Nature Restoration Act is an incentive for its recovery, although it is true that many measures need to be coordinated with this policy to obtain truly positive results. The reduction in good agricultural practices, currently under debate, generates uncertainty about this objective. However, decision-makers should not forget that the new regulation offers some substantial flexibility to member states to achieve nature restoration objectives. It allows countries to develop and implement their nature restoration plans according to their own national context, to be worked on at the national level with transparency and social participation.

Bearing in mind that the recovery of the Mar Menor requires, among other things, a substantial environmental reconversion of the agricultural production model of the Campo de Cartagena, the relaxation of agri-environmental measures must be done in coordination and with the collaboration of all the agents involved, including, of course, the agricultural sector. To this end, the environmental reduction of agricultural activity must be scientifically evaluated, if it does not hinder the reduction of agricultural pollution at source, the application of nature-based solutions and the recovery and extension of natural wetlands. In short, the backtracking on agri-environmental measures should be treated with caution, otherwise, it may make it difficult for the Nature Restoration Act to be truly effective in the Mar Menor.

Notes

1 The Mar Menor is on the list of Wetlands of International Importance (RAMSAR) and Specially Protected Areas of Mediterranean Importance (SPAMI). It has been declared

a Protected Landscape of the Open Spaces and Islands of the Mar Menor, a Regional Park of Salinas and Arenales de San Pedro del Pinatar, a Site of Community Importance (SCI) "Mar Menor" and a Special Protection Area for Birds (SPA) "Mar Menor". It has also been declared a European zone vulnerable to nitrate pollution since 2001 (Directive 91/676/EC) and a sensitive area (Directive 91/271/EC).

2 Article 1.2 of the Nature Restoration Act.
3 https://www.consilium.europa.eu/es/press/press-releases/2024/03/26/support-for-farmers-council-endorses-targeted-review-of-the-common-agricultural-policy/.
4 Recital 1 of the Nature Restoration Act.
5 Recital 4 of the Nature Restoration Act.
6 Recital 7 of the Nature Restoration Act.
7 Communication from the Commission. The European Green Deal COM (2019) 640 final.
8 Recital 14 of the Nature Restoration Act.
9 Recital 28 of the Nature Restoration Act.
10 Recital 83 of the Nature Restoration Act.
11 Directive 2008/56/EC of the European Parliament and of the Council of 17 June 2008 establishing a Framework for Community Action in the field of Marine Environmental Policy (Marine Strategy Framework Directive).
12 The drawing up of national restoration plans should not entail an obligation for member states to reprogramme any funding under the CAP, CFP or other agricultural and fisheries funding programmes or instruments under the multiannual financial framework 2021–2027 to implement this Regulation (Recitals 78 and 79 of the Nature Restoration Act).
13 Recital 20 of the Nature Restoration Act.
14 Recital 54 of the Nature Restoration Act.
15 Article 14.12 of the Nature Restoration Act.
16 Recital 55 and 56 of the Nature Restoration Act.
17 Recital 57 of the Nature Restoration Act.
18 https://www.laverdad.es/murcia/docena-planes-regionales-mar-menor-siguen-pendientes-20240229012809-nt.html#vtm_funnel=exito-login-gis&vtm_tipoProceso=gis&vtm_procesoFinalizado=si&vtm_proceso=login-gis&vtm_tipoRegistroLogin=login-gis.
19 Recital 58 of the Nature Restoration Act.
20 Regulation (EU) 2021/2115 of the European Parliament and of the Council of 2 December 2021 laying down rules on support for strategic plans to be drawn up by the member states under the common agricultural policy (CAP strategic plans), financed by the European Agricultural Guarantee Fund (EAGF) and the European Agricultural Fund for Rural Development (EAFRD), and repealing Regulations (EU) No. 1305/2013 and (EU) No. 1307/2013.
21 https://porotrapac.org.
22 https://www.carrodecombate.com/2024/02/27/no-podemos-seguir-asi-algunas-claves-sobre-las-protestas-de-los-agricultores/.
23 More than 335 organisations signed the letter.
24 https://eeb.org/wp-content/uploads/2024/03/Joint-letter-to-the-EU-Commission-to-reconsider-the-loosening-of-the-CAPs-green-architecture.pdf.
25 Proposal for a Regulation of the European Parliament and of the Council amending Regulations (EU) 2021/2115 and (EU) 2021/2116 as regards standards for good agricultural and environmental condition, climate, environment and animal welfare schemes, modifications to CAP strategic plans, review of CAP strategic plans and exemptions from controls and sanctions (COM (2024) 139 final 2024/0073 (COD)).
26 https://www.mapa.gob.es/es/pac/pac-2023-2027/.

References

Cabello, V., Brugnach, M. (2023). Whose waters, whose nutrients? Knowledge, uncertainty, and controversy over eutrophication in the Mar Menor. *Ambio*, 52 (6), 1112–1124. DOI: 10.1007/s13280-023-01846-z.

Carreño, M.F. (2015). *Seguimiento de los Cambios de Usos y su Influencia en las Comunidades y Hábitats Naturales en la Cuenca del Mar Menor, 1988-2009, con el Uso de SIG y Teledetección*. Tesis Doctoral. Universidad de Murcia.

Comité de Asesoramiento Científico del Mar Menor. (2017). *Informe integral sobre el estado ecológico del Mar Menor*. https://canalmarmenor.carm.es/

Esteve Selma, M.A., Martínez Martínez, J., Fitz, C., Robledano, F., Martínez Paz, J.M., Carreño, M.F., Guaita, N., Martínez López, J., Miñano, J. (2016). Conflictos ambientales derivados de la intensificación de los usos en la cuenca del Mar Menor: una aproximación interdisciplinar. 79–112. En León, V.M., Bellido, J.M. (eds.) *Mar Menor: una laguna singular y sensible. Evaluación científica de su estado*. Madrid: Instituto Español de Oceanografía, Ministerio de Economía y Competitividad. Temas de Oceanografía.

Giménez Casalduero, M. (2022). El Mar Menor y la contaminación por nitratos: nuevos instrumentos jurídicos, misma incertidumbre. *Actualidad Jurídica Ambiental*, (128), 11–41.

Giménez Casalduero, M., Pedreño Cánovas, A., Ramírez Melgarejo, A.J. (2021). Vulnerabilidad ambiental y "alimentos baratos": los límites del Derecho ante la industria porcina. *Revista Aranzadi de derecho ambiental*, (48), 203–229.

Gutiérrez Llamas, A., Giménez Casalduero, M. (2009). La ordenación, planificación y gestión del litoral en la Región de Murcia. En García Pérez, M. (coord.), Sanz Larruga, F.J. (dir.) *Estudios sobre la ordenación, panificación y gestión del litoral: hacia un modelo integrado y sostenible*. A Coruña: Fundación Pedro Barrié de la Maza. 299–312.

Martínez, J., Esteve, M.A. (2019). *El colapso ecológico de la laguna del Mar Menor*. Informe OPPA 2019. Retos de la planificación y gestión del agua en España. 61–70. https://fnca.eu/biblioteca-del-agua/documentos/documentos/Informe%20OPPA%202019.pdf

Martínez, J., Esteve, M.A., Guaita, N. (2017). *La crisis eutrófica del Mar Menor. Situación y propuestas*. Informe OPPA 2017. Retos de la planificación y gestión del agua en España. 130–139. https://bit.ly/InformeOPPA-2017.

Ruiz Fernández, J.M., León, V.M., Marín Guirao, L., Giménez Casalduero, F., Alvárez Rogel, J., Esteve Selma, M.A., Gómez Cerezo, R., Robledano Aymerich, F., González Barberá, G., Martínez Fernández, J. (2019). *Informe de síntesis sobre el estado actual del Mar Menor y sus causas en relación a los contenidos de nutrientes*. https://fnca.eu/biblioteca-delagua/directorio/file/2897?search=1.

11 La Manga del Mar Menor. Architecture for environmental colonisation[1]

Miguel Mesa del Castillo Clavel

Development with the comparison slider widgets

A comparison widget is an interactive graphic component used in website user interfaces to help users understand the time evolution of a given scenario. It is now a very common resource in journalistic investigation and advertising which has superseded the classic before/after photographs used for, say, beauty treatments or the developmental expansion of a territory, as can be observed in the illustration from the excelent article published in the newspaper *La Verdad* in August 2023 (López Hernández, López Zumel, & García Bastida, 2023). Sliding the divider right and left allows us to compare two static images taken at different times and superimposed. It appears first as an objective resource which allows us to observe two realities and compare them comfortably to identify and verify the effects of the passing of time. However, if we pay attention, the 'narrative' effect of this graphic tool is not as objective as it seems, let alone neutral. As any image or group of images, this one is also soaked up in ideology.[2]

The question is what happens with the divider? What is there between the two superimposing images? The animation suggests that we are watching an animated sequence in which by sliding the divider we can choose at what time in the occupation of the territory we want to stop time. However, the two images are merely static, they are two photographs about which we can only choose on which one to land, but we cannot fill the temporal gaps with other intermediate images.

This highly effective visual trick (a 'before and after', as those orthodontics and acne treatment adverts which, incidentally, do not show the long months of suffering and the hours in the dentist's chair in between the two photographs) is also used to explain why nostalgia is often shown associated to forms of essentialist environmentalism. The question is what is left between the first and the second images, that the divider seems to wipe, windscreen-style, but in reverse?

We could accompany Bruno Latour in saying that between the two images, there is a 'black box',[3] a set of political proceedings, of debates, ideological constructions, material productions, standards, etc. which are hidden between the 'beginning and the end' of a particular closed-up fact or object, whose internal functioning escapes our understanding but which we take for granted as de facto.

DOI: 10.4324/9781003489078-12

In my opinion, the black box covered by the divider is a symptom of the way in which the recent historiography of the Modern Movement in Spain has discharged some of the political load, environmental complexity and ideology from the work of the architects of the second Spanish modernity, when they put in place the plans for modernisation and economic activation of the countryside and the coast for Franco's Spain after the civil war.

The colonisation of the dunes

An example of the above is the exhibition entitled *La colonización de las dunas* [*The colonisation of the dunes*], organised in 2021 by the Colegio de Arquitectos of the Murcia region, in collaboration with the Docomomo Ibérico Foundation[4] and the Universidad Politécnica de Cartagena, as part of the Conference *Arquitectura y Medio: el Mediterráneo* [*Architecture and Medium: the Mediterranean*]. In my opinion, the title could not have been more appropriate because the process of extractive occupation of the tourist industry in the Mediterranean is very similar to a colonial expansion plan. Iain Chambers and Marta Cariello have explained how the Mediterranean has been the subject of a form of colonisation which understands the so-called *colonial periphery* as an irrelevant part of the conceptual production of the Mediterranean as understood nowadays (i.e. as a part of modern Europe). As a consequence, this form of colonisation has granted privileges to the identity of the northern shores to the detriment to Asian and African ones, thus simplifying and reducing the diversity of languages, ways of living and temporalities of the Mediterranean to a single idea of what the Mediterranean is (Chambers & Cariello, 2019). In this sense, as we will see later, the urbanization project for La Manga del Mar Menor fits perfectly with what is expected of a tourist stay in the Mediterranean: enjoying the good climate, clean water, tasty and healthy eating and the ancestral culture of fishing, salt lakes and even bullfighting, which had such a presence in his project. Obviously, the title of the exhibition, the phrasing 'colonisation of the dunes' hides a 'civilising' and dominating intention so that the only possible Mediterranean is that of colonial Europe.

La Manga that could have been but was not

The exhibition displayed a large eight-metre-long model *representing* the project designed by the famous Catalan architect and urban planner Antonio Bonet Castellana, in collaboration with the Josep Puig-Torné, also an architect. The main aim of the exhibition was to draw attention to Antonio Bonet's legacy in the Murcia region and to recognise his work in La Manga by showing the virtues of his 'colonising' project and denouncing the misrepresentation and misuse of his ideas, which caused the development and landscaping disaster that – for the curators – La Manga is now. The title *La Manga that could have been but was not*, which appeared time and again in articles, interviews and news items,[5] summarises nostalgically the motivations underpinning the exhibition. The main argument is basically one that has been present for a long time in the local media and is the narrative that has become popular: La Manga had a good architectural project and an excellent development plan but the predatory greed of the property development sector and mass tourism thwarted them. Thus, democratising it and making it accessible for leisurely holidays was to blame for the tasteless, vulgar and deregulated production of its planning, and the catastrophic and senseless destruction of its environmental and landscape heritage.

Nevertheless, all these statements, which suit an environmental grief process ('What could have been but was not', 'La Manga in the past') leave many elements out in an incomplete narrative (as the slider widget does effectively), because La Manga has not suffered from a geological phenomenon nor the unavoidable fall of an asteroid. Antonio Bonet's La Manga, as stated in the exhibition, is an *exquisite* colonisation project of an absolutely unique place, with immensurable landscaping and environmental value and the architecture, very irresponsibly, contributed to its devastation.

Bonet and Puig-Torné's project comprised a series of 12 high-density cores. Each core boasted a 21-storey tower with its corresponding quays and artificial island, as well as a large marina, retail and leisure areas, and a church. Furthermore, there was serious consideration about building an airport and a bullring, amongst other facilities and of course, a nine-hole golf course – everything a resident needs within 500 m – as the project's authors claimed in the design report. It must be pointed out that, without describing the project in detail, and in contrast to what is normally argued (that the dunes were spared, and all the density was built around the tower cores), there was the intention of building up and privatising practically the whole area from the very first drawings.

If one looks at Bonet's original drawings in detail (Figure 11.1), there is evidence that, in contraposition to what is stated in the project's design report, the intermediate areas between the cores would be developed, and the dunes would be divided into plots on which to build detached and terraced houses In fact, this is fundamentally rather similar to what is there nowadays, although then it was destined only for the middle and upper classes, and not the working class, which at that time simply did not go away for the summer.[6]

An extraordinarily talented architect, Bonet was Le Corbusier's student and collaborator, and his work was widely recognised. A republic sympathiser, he had to leave Spain for Argentina, from where he developed similar projects to those in La Manga, such as Punta Ballena in Uruguay. Returning from exile, Bonet devoted himself to one of the Francoist projects with greater transformative and *modernising* projects in Europe, one that placed tourism at the centre of Spain's financial and industrial development. When, under Manuel Fraga Iribarne, Minister for Information and Tourism, La Manga was declared Centre of National Tourist Interest, it was the death knell for the landscape and the ecosystem of La Manga, and the beginning of a series of environmental abuse yet to be concluded.[7]

With its new planning classification, La Manga and therefore the Mar Menor ceased to be a unique landscape and ecosystem in the European Mediterranean and became available land to be exploited. As in all colonisation processes, the land, including its people and its multiple life forms, could now be looted and deprived of its singularities.

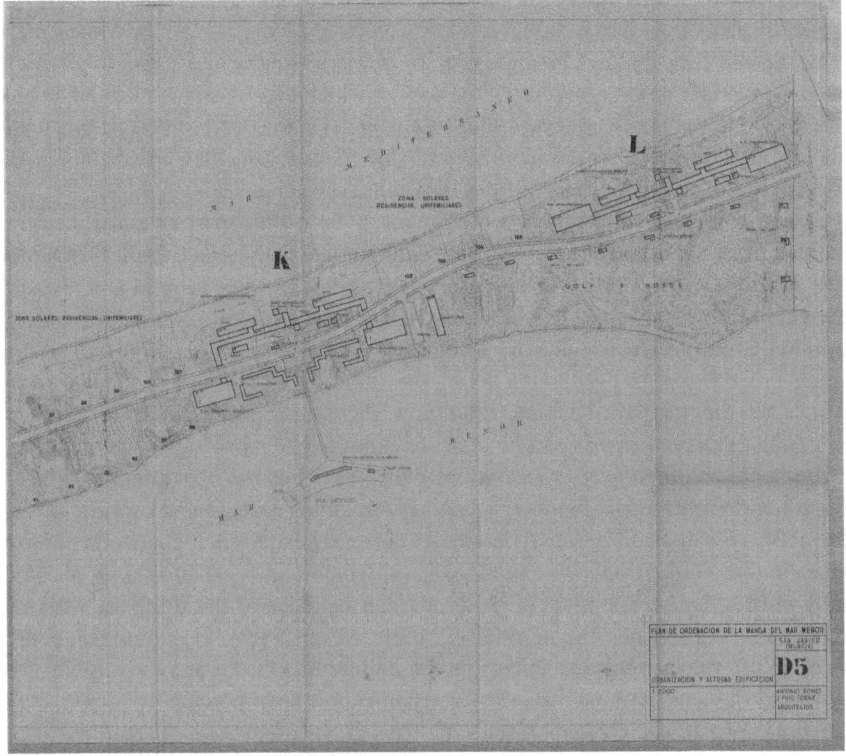

Figure 11.1 Original drawing of a sector of the La Manga del Mar Menor Development Plan, Antonio Bonet and Josep Puig-Torné.

In the colonial logic, the colonised are the beneficiaries, they are not submitted. One only needs to remember the current debates in Spain about the legitimacy of the so-called *Black Legend* which, for some revisionist historians close to nationalist positions, was a political invention by rival powers. By distorting the facts and with rather outlandish arguments which, by the way, have been duly contested by some specialists (Strahele, 2021), according to this alternative Spanish history, the commonly acknowledged predatory offensive of the *conquistadores* brought many more benefits to Latin America than negative effects. Similarly, the narrative about La Manga's colonisation (*A Paradise between two seas* was the constant mantra) contains implicitly the 'advantages' that the operation of occupation and devastation of its dunes ought to bring to the still 'wild' territory which still had to be civilised or developed.

Ramón del Castillo (del Castillo 2021, p. 109) has highlighted the work of the art critic Robert Hughes on the history of the United States to explain the features of North American culture of the late 20th century (Hughes, 1994) and the importance of the latter attaches to the connection of the pioneers with the landscape in the configuration of the dominating narrative about the American epic story. Hughes argued that for the majority of American colonisers, the landscape was never an object of aesthetical satisfaction, bar a few exceptions – and always for white Americans (del Castillo, 2021). The logic used by modern architecture critics to justify its 'colonising' is a strong reminder of the way in which the American pioneers related to the landscape: landscape was not landscape [...] it was land, property, raw material. Its qualities were practical: fertility of soil, the nutrients they could contain, the availability of water, the type of trees it sustained (Hughes 1994, p. 142).

For pioneers, for Bonet and eventually for modern architecture, landscape is not tackled from a spiritual or aesthetic but from the stance of extractivism: 'What matters is not how can I preserve it, understand its dynamics or rebuild and preserve the cultural forms inserted in it but how can I extract a benefit from it?' This, in my opinion, does not always have to mean a series of aggression or processes of pollution and destruction. Without a doubt, the saltwater lagoon, the channels and the canals and the traditional fishing techniques, such as the *encañizadas* (reef barriers to trap fish) and *paranzas* (reef boxes), are extractivist technologies, but they were also the origin of many transformations which enriched the ecosystem of the Mar Menor and promoted its biodiversity. A very notable example is the *pinna nobilis*, the noble pen shell, an oyster-type species currently experiencing an endemic situation in the whole of the Mediterranean and which had been introduced accidentally in the Mar Menor in the 1970s, as a consequence of the dredging works in the Estacio canal when the marina was built. Nevertheless, the noble pen shell's work of filtering out the nutrients from the ravines and the Quaternary aquifer was key until the successive episodes of eutrophication have practically eliminated it (Instituto Español de Oceanografía, s. f.). To a great extent then, the equilibrium of the ecosystem in the lagoon depends on the participation, often spontaneous and unexpected, of more-than-human entities and technologies.[8]

Architecture for colonisation

However, in La Manga, we do not find these multi-species collaborative adaptation processes in the line of the studies carried out by Anna Tsing (2021) or of the recent work by María Puig de la Bellacasa on soil (2023), for instance. This is an unprecedented environmental disaster caused by a territorial colonisation process associated with a political project (first, the industrial, financial and social transformation of the state under Franco and then the several land acts, property bubbles and supremacy awarded to bricks and mortar). It should not be forgotten, as explained by Lino Camprubí, that many Spanish architects worked for the National Institute of Colonisation (Instituto Nacional de Colonización, s. f.) designing the projects for the new populations which would be the backbone of the Spanish territory by improving the agricultural sector productivity as one of the main strategies underpinning the autocratic plans of post-Spanish Civil War Francoism (Camprubí, 2017). Over 300 new colonising towns caused the displacement and submission of 55,000 families to the land because of the commitments associated with the allocation of land and properties. Engineering, science, technology and, in our case, architecture were not only an instrument of power but they also added purpose to the Francoist regime (Camprubí, 2017). The agricultural engineers responsible for the National Institute of Colonisation, Ángel Zorrilla Dorronsorro, Professor of Political Economics at the Escuela Especial de Ingenieros Agrónomos (Agricultural Engineers School) in Madrid and Emilio Gómez Ayau, Founder of the Centro de Estudios Agrosociales (Centre for Agrarian and Social Studies), were in charge of putting into practice some of the ideas inherited from the Spanish regenerationist Falangism, heir to the agrarian and social Catholicism which, duly adapted, offered a good strategy to drive the agricultural production in the country and improved the conditions of the rural sector. In order to remove themselves from the anarchist experiences of expropriation and re-distribution before the Spanish Civil War, it was necessary to conceive an ideologically neutral colonising plan and technocratic profile led by engineers and architects (Camprubí, 2017, p. 67), a supposedly 'neutral' vocation which also contaminated the plans for tourist development in the Centres of National Touristic Interest. Most of the colonising towns were designed by already prestigious architects, such as Fernández del Amo, de la Sota and Fisac. Many of the most famous Spanish architects of the time, such as Francisco Javier Sáenz de Oíza and Alejandro de la Sota, designed projects for La Manga. In addition to those of Bonet, along the coastline buildings by Javier Carvajal, Juan Daniel Fullaondo, Juan Antonio Corrales and Ramón Vázquez Molezún, Miguel Fisac, Fernando Garrido and Joaquín Sebares, to name but a few,[9] have been preserved with some degree of alteration.

Although very distinct processes, the 'colonisation' of the dunes and the rural areas share many similarities in terms of productivity. In any case, the key differences only depend on the socio-technical programme which motivated it (the promotion of tourism or the agricultural sector). In territorial terms (and evidently, not in terms of social justice and the balanced share of the resources), there is not much difference if the occupation of the dunes was intended to accommodate tourists with a high purchasing power or if the countryside was destined for the working classes, who had been impoverished by the war. In both cases, this is a project of submission and extractivism, as with any colonial process.

It is symptomatic that to this day, when trying to devolve its 'rights'[10] to the Mar Menor lagoon, a difficult to solve contradiction immediately emerges: the discourse by those who defend the submission of human interests to the 'laws of nature' and those who (such as Franco's Ministry of Information and Tourism) proposed to submit nature to human needs has the same vocabulary of domination and submission, perhaps ignoring that it should have been absolutely necessary, before deciding which side of the debate we are on, to discuss what we mean by 'nature'. In the words of the geographer Erik Swyngedouw, who suggested the field of urban political ecology, 'Nature does not exist! Sustainability as symptom of a depoliticized planning' (Swyngedouw, 2011).

Environmental mourning

Experimenting with the loss of the landscapes of the Mar Menor and La Manga has placed the coastal populations into a sort of environmental grief. On the one hand, it is manifested in a fatalistic way as an unstoppable transformation (submission of nature); on the other hand, it has driven a grieving process which demands a very problematic nostalgic restitution of the landscape and the ecosystems before the mass 'colonisation' of the coast (the restitution of a primitive Arcadian Nature). As Timothy Morton says, 'One damaged concept is "Nature" – I capitalise it to denature it – damaged and damaging, almost useless for developing ecological culture' (Morton, 2010). And Ramón del Castillo demands the same in a very practical and somewhat provocative way, when he says that protecting a landscape should be the duty of the ministry of culture, not of the environment (del Castillo, 2021, p. 114).

For del Castillo, when aiming to restore the past in nature, as it may be the case of La Manga, what we are really aiming to do is to restore *ideas* about the past (del Castillo, 2021, p. 115), because nature is a human invention and therefore, to restore it would be to reconstruct the ideas which made up that invention at a given time in history – something which is objectively very difficult to do. Landscape is a cultural *construct.* It is not possible to think of a primitive, original, Arcadian landscape because the landscape is always a transformed nature, a *denaturalisation* of the land.

It seems impossible to escape from the obligation to choose between the classics – John Muir (ecocentric conservationism) or Gifford Pinchot (anthropocentric conservationism)[11] – and to endorse the polarisation between the social and natural realms. But as explained by Latour, Serres (1992), Haraway (2016), Tsing, Despret (2015), Morton and others (with notable differences albeit with notable fundamental coincidences), nature is not a larder, an outside laid out to be exploited, as it used to be claimed, and is still defended by modern architecture and sustainability culture, it is not an Arcadian Eden to which one can return, but a symmetric and unstable composition of entities of very diverse scale, category and origin. It is possible that, as pointed out by Alain Roger in his short treaty on landscape, ecology is a mire of biology and theology (Roger, 2007, p. 174). But to examine the Mar Menor and la Manga, nowadays a different gaze is needed to the one offered by the slider in the article mentioned at the beginning of this essay, a gaze that gives us access to the possibility of an architecture entwined with nature; a new scenario in which the relationship between architecture and nature is not that of figure-ground but of assembly. An architecture which is attentive of the relationships of mutualism with other species, that is interwoven with a nature that cannot be understood any longer as a harmonious self-centred whole and separate from the processes that produce it, from the social, environmental, technological, architectural and developmental conflicts, and of the policies that make it up. An architecture which is attentive to the landscape in its potentialities and complexity, rather than in its nostalgic longing for an irrecoverable and idealised natural past.

Conclusions

Bonet and Puig-Torné's project for La Manga was a milestone in the history of development in Spain and its architectural quality has been sufficiently recognised by the specialised critics. In the creative biographies of both authors, La Manga and some of the buildings designed and built by them there occupy a prominent place. This essay does not intend to be a critical analysis of Bonet's work nor does it claim to judge the validity of his creations per se, but to warn about the consequences for the Mar Menor landscape and ecosystem of a project whose main drive was to ensure the continuity of a colonising process with totalitarian tinges to promote the tourist industry, in the same way that agriculture had been pushed by the National Institute of Colonisation.

The environmental grief experienced by many people affected by the ecological and landscape disaster in La Manga and the Mar Menor has led to the emergence of two very different (even contradictory) ways of coping with loss, on the one hand, a resigned acceptance of the transformation of the environment, which is recognised in the idea of a Nature submitted to human needs, and, on the other hand, a longing to restore the ancient beauty of the place and the enormous vitality of its ecosystems prior to the process of massive 'colonisation' of the coast. But if the idea of Nature is itself a human invention, a cultural construct that evolves over time, the idea of an original landscape is a näive fiction, since the landscape is transformed by multiple technologies and extractive processes, such as fishing or salt mining.

Thinking about the colonising project of La Manga also gives us access to rethink our relationship with an idealised and irretrievable natural past. Bonet and Puig-Torné's development project in La Manga kick-started the long process of territorial occupation and landscape alteration which has not finished yet.[12] This process, still ongoing, requires an urgent and in-depth reflection about what the loss means for the coastal populations, about how, in order to restore an ecosystem which stirs feelings of nostalgic evocation, a gradual 'decolonisation' by architecture, planning and ecology disciplines is required.

Notes

1 This paper has benefited from the support of the following institutions: Ministry of Science and Innovation, through the research project 'Crisis climática, salud mental y bienestar en el Antropoceno. Una aproximación desde la ontología histórica' (PID2021-124477OA-I00 funded by MCIN/AEI/10.13039/501100011033, by 'ERDF A way of making Europe').

2 As stated by philosopher and art historian Georges Didi-Huberman, a document encompasses at least two truths, the first of which is always insufficient (Didi-Huberman, 2015). Didi-Huberman has examined in-depth the political character of the images and their interpretation in several of his works.

3 Latour, by studying what goes on inside scientific laboratories, develops this idea in many of his writings, for example: *Give me a laboratory and I will move the world.* In K. Knorr et M. Mulkay (editors) Science Observed, Sage, 1983, pp. 141–170 [New edition slightly abridged in Mario Biagioli (editor) Science Studies Reader, London Routledge, 1999].

4 Docomomo Ibérico is an organisation committed to the defence and appreciation of the architecture of the Modern Movement in Spain and Portugal.

5 The archives hold abundant material on the subject. See, for example: Fernández (2021) and Ruiz Salmerón (2023).

6 It must be clarified that, although there was evidence of urban deregulation following the sale of land to speculators, payments to contractors with plots and permits awarded with increased buildability throughout La Manga's history of 'colonisation' as well as the occupation and fencing off green areas, the truth is that the original project already prepared the land for what would happen later. It would not be fair to exonerate the responsibility for the current situation to the whole Francoist political, financial, technical and propaganda-spreading machine which had a notable ally in Bonet and Puig-Torné's architecture.

7 There are many similarities between the places affected by the Act of Centres and Areas with National Touristic Interest of 1962, such as Puerto Banús, Torremolinos, Santa Pola and Marbella. But the case of La Manga has very particular characteristics. Hacienda de La Manga de San Javier and Hacienda de La Manga de Cartagena, declared Centres and Areas with National Touristic Interest in 1969 and 1966 respectively, had an impact on one of the most extraordinary landscapes along the Spanish Mediterranean coast. The declaration and the ensuing series of catastrophic events provoked an enormous environmental damage: lakes drying, loss of dunes surface, degradation of biodiversity, amongst many other disasters. For a detailed study of the impact of the Act on La Manga, see García Ayllón Veintimilla (2013).

8 When considering the agency of more-than-human entities, Bruno Latour encouraged us to reflect on the complex interactions between humans and non-humans in the configuration of the social reality. This implies recognising that non-human entities have *agency*, that is, they can have interests, objectives and effects that must be taken into account (Latour, 2008).

9 Some of these buildings have been listed and protected by the Fundación Docomomo Ibérico (Documentation and Conservation of buildings, sites and neighbourhoods of the Modern Movement) (Fundación Docomomo Ibérico - Arquitectura moderna España y Portugal, 2024).

10 The legal personality of the Mar Menor was officially recognised on 28 September 2021 with the passing of Act 11/2021, dated 9 September, for the protection and integral recovery of the Mar Menor. This act set a significant milestone in the conservation and sustainable management of this ecosystem, granting it more solid legal protection and recognising its environmental and social importance.

11 It could be argued that the ecocentric conservationism, represented by John Muir (1838–1914), and the anthropocentric conservationism, represented by Gifford Pinchot (1865–1946), are the two great American traditions dealing with the relationship between human beings and nature. These two perspectives have had a significant impact on the way the environment in the United States and the world is managed and conserved.

12 Still today, a week before finishing the writing of this essay, there is the disastrous news about the restart of the construction project of a pedestrian walkway connecting the Salinas de San Pedro natural park with the last developed section of La Manga (*La Manga resucita el proyecto de pasarela peatonal con las salinas de San Pedro | La Verdad*, s. f.).

References

Camprubí, L. (2017). Los *ingenieros de Franco: Ciencia, catolicismo y Guerra Fría en el Estado franquista*. Crítica.

Catálogo de datos del IEO - Instituto Español de Oceanografía. (2022). Retrieved 21 February 2024, http://datos.ieo.es/geonetwork/srv/spa/catalog.search#/metadata/ffbe0f01-cb71-4471-bfc1-a22b78d8ff75.

Chambers, I., & Cariello, M. (2019). *La questione mediterranea. Mondadori.*

del Castillo, R. (2021). *El jardín de los delirios: Las ilusiones del naturalismo*. Turner.

Despret, V. (2015). *Habitar como un pájaro: Modos de hacer y de pensar los territorios*. Cactus.

Didi-Huberman, G. (2015). *Cuando las imágenes toman posición*. Antonio Machado Libros.

Fernández, C. (2021). La exposición que la Región le «debía» a Bonet Castellana, el arquitecto que soñó con otra Manga del Mar Menor. *Murcia Plaza*. https://murciaplaza.com/la-exposicion-que-la-region-le-debia-a-bonet-castellana.

Fundación Docomomo Ibérico—Arquitectura moderna España y Portugal. (2024, febrero 21). Fundación Docomomo Ibérico. https://docomomoiberico.com/.

García Ayllón Veintimilla, S. (2013). En los procesos de urbanización del litoral mediterráneo español, caso La Manga [Tesis doctoral, Editorial Universitat Politècnica de València]. En *Riunet*. https://doi.org/10.4995/Thesis/10251/28581.

Haraway, D. J. (2016). *Staying with the Trouble: Making Kin in the Chthulucene*. Duke University Press.

Hughes, R. (1994). La *cultura de la queja: Trifulcas norteamericanas*. Anagrama.

Instituto Nacional de Colonización. (2024). Retrieved 21 February 2024, de https://www.mapa.gob.es/en/ministerio/publicaciones-archivo-biblioteca/mediateca/colonizacion.aspx

La Manga resucita el proyecto de pasarela peatonal con las salinas de San Pedro | La Verdad. (2024). Retrieved 25 March 2024, de https://www.laverdad.es/murcia/manga-resucita-proyecto-pasarela-peatonal-salinas-san-20240317073246-nt.html.

Latour, B. (2008). *Reensamblar lo social: Una introducción a la teoría del actor-red*. Manantial.

Morton, T. (2010). Ecology as Text, Text as Ecology. Oxford Literary Review, 32(1), 1–17.

Puig de la Bellacasa, M. (2023). *El espíritu del suelo: por una comunidad más que humana*. Tercero incluido.

Roger, A. (2007). *Breve tratado del paisaje*. Biblioteca Nueva.

Ruiz Salmerón, F. (2023). De ecoturismo por el Mar Menor. Arquitectura medioambiental y racional de Bonet Castellana. *La Opinión de Murcia*. https://www.laopiniondemurcia.es/cultura/2023/08/24/arquitectura-medioambiental-racional-bonet-castellana-91263934.html.

Serres, M. (1992). *El contrato natural*. Anagrama.

Strahele, E. (2021). Melancolía Imperial y Leyenda Negra en el paisaje español actual. *Ecos imperiales: diálogos sobre la imperio nostalgia.Revista de Historia «Jerónimo Zurita», 99*. https://ifc.dpz.es/ojs/index.php/Zurita/issue/view/88.

Swyngedouw, E. (2011). ¡La naturaleza no existe! La sostenibilidad como síntoma de una planificación despolitizada / Nature does not exist! Sustainability as Symptom of a Depoliticized Planning. *Urban, 01*, Article 01.

Tsing, A. L. (2021). *La seta del fin del mundo: Sobre la posibilidad de vida en las ruinas capitalistas*. Capitán Swing Libros.

12 The assembly of the gaze

Composing wetlands in a glocal world

Mijo Miquel Bartual[1]

Introduction

Given the complexity of the present, constantly destabilized by pandemics, food crises, wars, or unprecedented developments in Artificial Intelligence, we find ourselves in the need to readjust our view of the world, even purposefully defocusing it or applying a wide-angle lens to include issues that we had considered peripheral until then. To illustrate this process, we have chosen a specific area, wetlands at risk. From among these territories, we have selected three specific projects: two linked to a freshwater lagoon located in the province of Valencia, the Albufera (GVA, 2022); and the third located in Murcia and linked this time to another saltwater lagoon, the Mar Menor. Both constitute very fragile ecosystems, with high anthropogenic pressure and a direct threat from the continued increase in sea level and temperature. In all three projects, mixed working methodologies (ranging from theoretical research to personal interviews, through role-playing games or artistic creation and other participatory practices) have been applied that attempt to help us integrate this complexity without reducing it to its basic lines.

In other chapters of this book, the organizers of the first two projects analyse their case studies in more detail. In my case, I have chosen to comment on them briefly as a participant in the first project of the Albufera de Valencia (Presentes Densos[2]) and as an observer of the Murcian project in which I collaborated punctually for one of its work sessions. This will allow us to better understand the adjustments we have made in the third project that we organize together with other colleagues, a project that we have called Agua Dulce and that works again on the Albufera de Valencia.

Presentes Densos

Through the experience of Presentes Densos, an annual research seminar on the climate emergency suffered by the Albufera Natural Park,[3] a working group was generated that combined the exploration of the territory with the shared reading of theoretical texts. This combined methodology proposed applying a magnifying glass view to a delimited space, putting the body in it through tours and processes of active listening to the different agents involved in it. In this case, the working

DOI: 10.4324/9781003489078-13

group met and listened to rice farmers, hunters, tourists, biologists, or inhabitants, all with a different relationship with the lake, composing a living fabric of experiences and variable tensions. Through their words, and for nine months, we understood that the survival of the Albufera involves a delicate exercise in negotiations on action schedules and chains of interactions between the different agents.

It is also evident that the central element around which many practices and conflicts revolve is the question of water, a problem that will only intensify in the future given the scarcity associated with the persistent drought in which we find ourselves. Similarly, the use of both pesticides and fertilizers in surrounding agricultural exploitations plays an important role in the eutrophication of the environment. This use is theoretically controlled by regulations following the different crises that have occurred since the early 1980s. However, in both territories, every time torrential rains occur, the water runoff produces immediate effects on the lagoons: we observe the image of massive death of fish and other forms of animal life, which we have almost normalized by force of seeing it. The accumulation of nitrates and phosphorus causes the proliferation of algae and anaerobic bacteria, which in turn causes the consumption of the present oxygen to the point where the fish die from anoxia. In the case of pesticides, massive uses also cause lethal effects in the ecosystem, cumulative over time. All this only increases tension in the dialogue between the different agents and shows that water management is going to be one of the central issues in our immediate future. In the case of the Albufera, there is also the added weight of population pressure and the industrial ring, almost more important than that of agro-industry, contrary to what happens in the Mar Menor. However, in both cases, there is the unlikely (or perhaps very truthful) lack of an updated management plan that realistically considers climate change.

In parallel, in the Presentes Densos working group, we shared our own knowledge as environmentalists, sociologists, artists, or technicians from the Administration, providing complementary information. This helped us draw a map in which legal regulations, incubation periods, or the appearance of species would all have to be balanced weights to keep the ecosystem afloat. But our bodies also registered a map of sensations that accounted for the sounds, textures, and changes in light in the environment. This close-up vision of the components of the working group gained depth through the reading of theorists such as Stengers (2005, 2015), Despret (2021), or Haraway (2016), who helped us detach from the specific data and try to project it into a broader context. The combination of information at different levels was a great success of this program, allowing us to walk through theory, discuss life stories, and linked them to abstract sensations.

Haraway (2016) taught us that we live in a dense present in which the level of complexity multiplies exponentially when we try to "unravel" the causes and propose solutions to situations of ecological harm. We cannot dream of returning to a theoretical primordial state in which "Nature recovers" because we cannot even conceptually maintain that separation. Nor can we expect to overcome conflicts and their concatenations through agreed radical solutions. If anything, we learned from listening to the different agents of the territory that they usually do not coincide in the diagnosis nor in proposals for action. It is our task then, according to

Haraway: "to become capable of responding reciprocally to each dense present". As Bruno Latour (2011, 2017) indicates, "Nature is no longer what is encompassed from a distant point of view from which the observer can ideally jump to see things 'as a whole', but the assembly of contradictory entities that must be composed as a whole". It is not about discovering the solution to a defined problem but about composing a reality diplomatically through the assembly of different perspectives. The three projects have tried to account for this purpose in different ways and in my opinion, with equally diverse results.

In this first program, we did not openly address this objective but rather let it hover among the participants, without posing the need for its crystallization. Although the program extended over nine months, the participants felt very comfortable, almost like high-culture consumers who come out of the experience refined in their beliefs but without any need to react or make any return to the territory where they had been wandering. We "stay with the trouble", without getting involved in it, happy to have met and exchanged knowledge and experiences, but we did not become a learning community or feel mobilized to action. We were rather witnesses to the development of a controlled tragedy where the victim would ultimately be the territory itself. It is true that the program was presented from an art centre and that is inherently demobilizing, although it offered an excellent platform to share all the information accumulated throughout the program. It is noteworthy that the communicative effort made since the centre not only collected the sessions and texts of the participants but also generated a permanent archive with many other related materials.

We might say that one of the elements that kept a distance between us and the Albufera, in our bubble of expertise, was the absence of meeting spaces among different agents. It was us who listened to one story after another as they came to our encounter, which at the same time granted the necessary space and time to understand each of the agents. However, the fact that we did not listen to how those stories were woven together gave us a basic vocabulary but not a structure of enunciation. Another factor that prevented us from becoming a learning community was the fact that there was no space to agree on a final narrative, albeit contradictory or incoherent, if necessary, that tried to overcome the complexities and multifaceted visions, to offer some shared assembly, neither among the agents involved nor even among one another.

The case of the Mar Menor

In the case of the Mar Menor, we analysed the confluence of two projects that end up merged: *Shared Dialogues*,[4] and *Tools for Citizen Participation in Ecosocial Transition Processes*.[5] Briefly, I will comment on processes and methodologies in order to extract the elements that we consider fundamental when applying them in the research of the Albufera.

Shared Dialogues is a transdisciplinary project that arises from the collaboration between artists and scientists based on two-year participatory research in the Mar Menor and Campo de Cartagena, which continues to this day in a third phase that will last until 2025. Initially, during the first year, the experiences of 28 people from the entire territory regarding the transformations that had occurred there in the last 50 years were collected. This allowed mapping agents as well as an initial investigation into the different ways of narrating what happened. In reference to this idea of assembly, three narratives were jointly written from different perspectives: those of farmers, activists, and people coming from the tourism sector. Subsequently, through discussion groups, negotiation was carried out around a common narrative, in which there were minimal agreements on the "truth" that the different agents were willing to assume and propose almost as "consensual fiction".

In a second phase, a dialogue group of between 8 and 12 people was created, including people from the Mar Menor and Campo de Cartagena with different positions on the lagoon crisis. In these meetings, various participatory methodologies were applied aimed at promoting dialogue with others, mutual knowledge, and joint reflection in a process of collective knowledge construction. Through the different sessions, not only stories were discussed but also the feelings associated with them such as the pain of losing the ecosystem or the discomfort of feeling blamed as responsible for the crisis. A generalized distrust towards public administrations for their inaction, and towards social movements for their "politicization", was also identified. The issue of personal/corporate responsibility in the ecological crisis of the Mar Menor was also addressed, verifying that personal perceptions of the impact of one's own activities differed greatly from one agent to another. However, unfortunately, what they mostly shared was the inability to think of a positive future for the Mar Menor.

Finally, the research team embarked on an artistic creation project through the design of 20 fictional visual stories based on previous co-research work. The drawings, narration, and structuring of the stories were generated in four blocks, extracting information from the different stories collected. The organization of information was divided into life stories (human and non-human); discourses, with their internal contradictions, dialoguing between different forms of knowledge; emotions surrounding an environmental catastrophe with the transition from sadness to action; and, finally, how to give voice and a future to the lagoon.

The different phases of these mediation processes were combined with a free course organized by the University of Murcia, called *Tools for Citizen Participation in Ecosocial Transition Processes*. To avoid repeating information, we will not detail the sessions, but the objective of this course was to empower citizens towards the necessary ecological transition of the Mar Menor (and, needless to say, the entire planet). Juan Manuel Zaragoza, its organizer, spoke of the need to address the problem not only from a technoscientific point of view but also including the cultural and social sphere. He also emphasized the need to address political issues, openly advocating for the opening of participatory processes that would foster mobilization and, therefore, the ability of citizens to influence public authorities.

Beyond admiring the precise articulation of the different elements and the communicative effort to share reflections and results, not only through the graphic project but also through infographics, stories, and conclusion sheets, the rigorous analysis of these projects has served us to prepare ours with a greater awareness of what we want to achieve. As a prior learning, it seems necessary for us to incorporate specific attention to emotions and the social issue into the project, often forgotten. As Latour says, we cannot separate facts from opinions but neither from feelings. Similarly, we are interested in exploring how to include the issue of responsibility without it turning into accusation or guilt, as well as the motivations that lead us to action.

However, the shared re-elaboration of narratives to agree on a story with which everyone can identify is a great success. A story is never a manifesto or a statement, a story always has an ambiguous relationship with truth. Likewise, the use of illustrations to tell the different stories and articulations makes the format seem more naive, less harmful, and therefore, more acceptable to all parties. Another of its great successes has been to open the door to the stories of the more than human, including the blue crab in its gallery of characters. Similarly, giving voice to the lagoon and dreaming of a possible optimistic future means that our imagination is not only blocked by sadness, resentment, and dystopian images, especially if practiced after precise knowledge of the environment.

The agency of citizens through the course of participatory exercises, and processes in which no separation is applied between researchers and researched in the construction of knowledge open the way for us and clear up doubts about how we can act on the issue of the Albufera, in Agua Dulce, a project that in a way is a continuation of what was learned throughout the other two projects.

Agua dulce

In December 2023, we received a small grant from the European Climate Foundation[6] to strengthen social movements advocating for the Albufera, through a process of accompaniment and conflict resolution, with the aim of uniting forces to achieve greater social and political impact. Therefore, we decided to form a multidisciplinary team composed of professionals from various fields: experts in natural space management, artists, professors from the Polytechnic University of Valencia, and specialists in citizen participation and group facilitation.

The objective was to bring together different perspectives on the Albufera through interviews and discussion groups to generate a collective narrative among participants. All the gathered information would serve as the basis for creating a comic book that would depict the collected material and be published by the end of 2024. The overall project was structured into five blocks, not necessarily consecutive in time: preliminary research of the context and mapping of relevant agents, facilitation process through interviews and discussion groups, creation of the collective narrative in comic form, return to the context, and presentation of results at a meeting between activists from the Albufera and the Mar Menor.

Phase I: preliminary research of the context and mapping of local agents of interest

Firstly, a research phase was conducted, involving a prior documentary and bibliographical review of the context and its history to ensure a rigorous and careful approach to the addressed context. Subsequently, documentation analysis was carried out, including institutional archives, media-produced archives, and archives produced by individuals, groups, and movements linked to territorial defence in Valencia, to contrast the narratives of different actors about the same territory. Secondly, a mapping of agents related to the project was carried out, a sociogram, and initial contacts were established to outline a possible group of individuals and/or groups interested in joining the overall process or some of its phases.

Phase II: interviews

This phase began to delve into the opinions and sensitivities of the participating agents regarding the topic at hand, aiming to draw a collective narrative through the diverse existing voices, with special emphasis on those that are generally less heard. To achieve this, we scheduled 10–12 semi-structured interviews with the agents selected in the previous research phase. After an initial delineation of the process, we prepared a preliminary script focusing on the current situation of the Albufera and the transformation it has undergone in recent decades, incorporating personal memories and emotional components that would serve as the basis for the collective narrative developed in block four. Theoretically, these interviews would also allow us to become aware of possible alliances and antagonisms to find common ground.

However, after conducting several interviews, we found ourselves in the position of convening an internal meeting due to doubts that had arisen regarding the Albufera agents we had selected, as well as concerning the script itself and the proposed discussion groups. Initially, we had prioritized the inclusion of different, if not antagonistic, positions, including agents with occasional or distant personal connections, such as nearby storage companies or other industrial park enterprises, as our interest was to achieve mutual listening among the different agents involved in territory management.

Nevertheless, at that moment, we considered the possibility of preferring interviews with people who had an emotional relationship with the Albufera since we believed it was worth prioritizing emotions and connections over attempting to comprehensively reflect all viewpoints. We concluded that many of the interviewees were already familiar with the discourses and reasons of the other parties involved in the governance of the Albufera and that perhaps it was not so much about facilitating once again a listening process between agents but rather about generating a mobilization process and visualizing different futures.

Furthermore, we detected fatigue, scepticism towards external discourses, and a sense of futility in these exchanges as when it comes to taking action, no changes in the different positions have been perceived. Often, the people serving as interlocutors remain the same as 20 years ago and have gone through successive territorial discussion groups, the creation of a Network of Entities, and many other attempts to generate agreements that, according to their opinion, have led nowhere. We believe that it is not true that these processes have been in vain since they have led to a recognition of the different discourses. Although it is considered that everyone seeks to protect their own interests in advance, it is acknowledged that everybody falls within logics understood by all parties. There are conflicts of interest but no blind misunderstanding among people. We cannot say the same about entities since the anonymity that characterizes them contributes to "dehumanize" their actions, as if the fact of acting under foreign structures eliminated the part of responsibility that could be involved. Therefore, we consider that the listening phase with the positive aspects it can bring has already occurred through other means.

Regarding the mobilization issue, we observe that the groups involved in citizen movements are in a phase of senescence within the natural cycle of any informal organization, a phase in which there has not been enough generational renewal. Therefore, they are rather inactive, expressing that feeling of fatigue and the absence of prospects for improvement, which logically affects their motivation. They emphasize that institutions are also not doing their part in implementing the measures proposed by neighbourhood movements and especially in finding a way to carry out a coherent transversal policy in which the different administrations involved converge (City Council, municipalities, Júcar Hydrographic Confederation, central government, etc.). This makes the Albufera remain "ungovernable".

For all these reasons, we wonder how to reactivate awareness of the current situation of the Albufera, which remains critical, to try to overcome the frustration and helplessness generated by a normalization of this state of affairs. From the interviewees, the need to involve young people to take the lead and perhaps find new ways to reactivate the different groups is openly discussed, so we find ourselves in the position of modifying our route. Regarding the scripts, we talk about reformulating them to focus on three issues: emotions, motivation to take action, and imagining the future.

Phase III: facilitation process

The project proposed the realization of four discussion groups with the different involved agents, where participatory listening methods would be facilitated with the aim of generating empathy through personal memories, capturing subjectivity and emotional connection with the territory. For this, collective dynamics were thought to be used (such as collaborative maps, drifts, and timelines), with an emphasis on the possibility of reaching a consensus among participating social agents on desired future scenarios for the community. Dialogue would allow finding a possible unifying common goal or, if this were not possible, contribute to weaving a network of connections. The information collected, together with the interviews conducted, would allow the construction of the collective narrative. At the time of writing this chapter, we are finishing the planned interviews and exchanging materials with the visual artist. But due to the reconsideration of the project, as we have already mentioned, we have questioned the usefulness of these groups and have decided to change the proposal again.

In that sense, instead of only holding discussion groups, we propose that at least a couple of sessions be transformed into a workshop taught by young people in one of the institutes in the area. In this way, we would like to promote that much-needed generational renewal mentioned by the groups. This activity would be carried out by young people from Extinction Rebellion together with a maximum group of 20 volunteers from a public institute in the Albufera. This would consist of a workshop on NVDA (Nonviolent Direct Action), as well as the joint design and planning of a collective action in the territory. Depending on the outcomes of the workshop and the response from the volunteers and institutes, it might be possible to establish a youth citizens' assembly proposing measures to address the crisis of the Albufera, akin to the one recently developed in Valencia.[7]

Phase IV: production of the collective narrative

In continuous dialogue with the facilitation process, we propose creating a collective visual story involving the local community. This format allows for a coherent return to the community, facilitating exchange among agents who do not always coincide in time and space, as well as the dissemination of the story beyond its production context to engage with different audiences. Therefore, the proposed process is of a coral nature based on individual experiences, which in no way seeks to reflect a single story but is built upon multiple experiences and positions, which may at times be contradictory, thus reflecting the plurality of a context like the Albufera. For this purpose, the methodology used is hybrid, intertwining tools of qualitative social science research with tools and supports of artistic practice. To this end, a visual artist, Miguel Brieva,[8] will collaborate in its development, basing his work on the information collected through interviews, workshops, and discussion groups. Regarding the comic's content, we had previously discussed the idea of framing it as in Murcia by Life Stories – Conflicts – Visions of the Future. We would like to illustrate two visions of the future, both in the realm of science fiction: one representing how we would all like the future of the lagoon to be, and the second portraying the future of the same lagoon if we do not take measures to address climate change. In any case, we propose generating the characters ourselves and structuring the dialogues so that no one feels singled out or may suffer the consequences of what is said.

Phase V: organization of a meeting of social movements from the Albufera/ Mar Menor

The first part of this phase would consist of identifying those agents willing to participate in this parallel meeting, knowing that a similar process is underway in the Mar Menor. The participation of social movements would be ensured in the meeting, as they would present the issues to their Murcian partners and vice versa. In this meeting, representatives of different groups should be able to exchange experiences and plan for continuity in creating a common identity (Wetlands Network) as well as the generation and exchange of shared tools for transition as, for example, the process to obtain the legal personality of the Mar Menor.

We would like to pay special attention to some issues, such as that this co-investigation process aims to address the connection with the territory and emotions, both positive and negative, since it is not just a problem of science and politics but also social (relational) and personal, though we are not seeking to romanticize it. Initially, we aim to weave a collective narrative in which different narratives and affections regarding a specific territory can coexist. Perhaps through this patient weaving, we can generate a shared capacity to respond to these present demands with something more than pure self-interest pragmatism, accessing a certain response-ability.[9] "Staying with the problem", in this case, would mean not dismissing any of the viewpoints in this dispute; it would mean coming together and feeling, thinking, and imagining until it is possible to generate a consensual fiction about what has happened and what is yet to come, reworking the narratives that currently polarize society about the Albufera.

All of these are issues that we will need to practice in order to mobilize our own knowledge so that we end up constituting ourselves as a learning community involved in what was initially just an object of study. This assembly work is especially necessary if we have to imagine that "we" of which we are part when assuming responsibility for the Anthropocene. All assemblies require intermediaries and/or facilitators so that collective engagement with the common cause that has been taken from us can be reactivated, according to Stengers (2015). But these intermediaries are not just us, it is the common cause (in this case, the Albufera) that has the power to bring us together and make us capable, but also, in a way, that obliges us to imagine a shared response full of nuances.

Given this state of affairs, the question we would like to ask would revolve around how to provoke a change in this unbalanced equilibrium, about what can move us to action: responsibility, motivation, consensual storytelling, indignation, anger, or the feeling of loss, of the breakdown of a liveable future. All of these feelings and thoughts are present when interviewees remind us that there is not even an updated Plan for the Management of Natural Resources in force, as it has been expired for several years. If there is no clear political will, they comment, what sense does it make to continue with talks and negotiations or with baseless propaganda claims. A good example of this is the government's publicized demand that the Albufera be declared a Biosphere Reserve, when it does not meet any of the requirements for it and there is no firm commitment behind it to make the necessary changes for it to be so. They tell you this with a trace of anger, and perhaps we should use those feelings as a motivation as well.

The climate crisis is a slow cataclysm: a matter of high politics rather than nature. Science and politics cannot be separated: biases, funding, specialized and therefore blind ways of seeing, and the absence of a collective exercise of thought rather lead us to think of *invisible third parties*, those who benefit from things not working. In this case, we could speak of a fundamentally extractivist capitalist inertia, which congratulates itself on things staying the same. For all these reasons, we advocate for a politicization of the problem, for a mobilization that allows for better protection of this territory and the lives that happen in it. Therefore, we propose to reread the project from another point of view that considers what elements move us to action, activating and socializing them.

The meeting sessions between activists from both lagoons are not merely occasions to share neutral knowledge but rather gatherings of individuals committed to provoking change, comparing the effectiveness of different tools to do so. We would like them to discuss how to construct another third party, preferably not originating from the university or the Administration, which highlights the need for a new social and institutional contract or, as Serres (1998) would say, a new natural contract that acknowledges that the environment is not just the background of our narratives but a central character in our story. We propose a sequence that progresses from our most negative emotions towards a moment of joy and could be described as follows:

1 Listening process: from a diverse sample of ecosystem agents selected according to the different presences (the density of their listening), thinking about how to listen to them, how to listen to ourselves, and how to listen to the territory.
2 Working on desire: desiring other presents together after interpreting the listening, translating their ideas into administrative language to streamline the processes, analysing infra-administration, forcing the territory beyond the legal and trusting without hope, opening ourselves to a new perspective, to a new narrative of the future.
3 Provoking the political moment of joy: desire is concretized in an action in its most political sense. The listening and interpretation process must stem from positioning ourselves in the struggle. For that, we must open ourselves to what was previously inconceivable in defence of the common cause. If we only discuss solutions, we leave it to others to ask the questions (Stengers, 2015), so let's dare to openly consider questions such as: What would a wetland union be? What pressure tool could it have? What would a territory strike look like? Or how to occupy spaces of power? Because the Administration is there to work on it, not to blindly admit its limits. But alone, we won't be able to do it; we must arm ourselves, articulate ourselves, and occupy spaces of power to give voice to the voice in the water; in the Albufera and in all the other water bodies.

Notes

1 This publication is part of the R&D&i project "Climate crisis, mental health and well-being in the Anthropocene. An approach from historical ontology (ANT-mentalhealth)" (PID2021-124477OA-I00), funded by MCIN/AEI/10.13039/501100011033/ and ERDF "A way of making Europe".
2 https://ivam.es/es/presentes-archivo/.
3 https://albufera.valencia.es/en.
4 https://www.shareddialogues.org/.
5 https://escuelalmargen.com/.
6 https://europeanclimate.org/.
7 www.asambleaclimaticavlc.com.
8 https://es.wikipedia.org/wiki/Miguel_Brieva.
9 Response-ability entails being fully present, not amidst idyllic pasts or salvific futures, but as mortal creatures entangled in countless unfinished configurations of places, times, matters, and meanings (Haraway, 2016).

References

Despret, V. (2021). *Living as a Bird.* Polity Press.

GVA (2022). *Parc Natural de l'Albufera.* Generalitat Valencia. https://parquesnaturales.gva.es/es/web/pn-l-albufera.

Haraway, D. (2016). *Staying with the Trouble: Making Kin in the Chthulucene.* Duke University Press.

Latour, B. (2011, November). Facing Gaia: Composing the common world through arts and politics [Lecture]. French Institute in London. Occasion of the launch of the Sciences Po Arts and Politics Program (SPEAP).

Latour, B. (2017). *Facing Gaia. Eight Lectures on the New Climate Regime.* Polity Press.

Serres, M. (1998). *The Natural Contract.* The University of Michigan.

Stengers, I. (2005). Introductory notes on an ecology of practices. *Cultural Studies Review, 11*(1), 183–196.

Stengers, I. (2015). *In Catastrophic Times.* Open Humanities Press.

13 Thick Presents. Stories for an ecosocial transition

Miguel Ángel Martínez[1]

Thick Presents

In one of her most recent books, *Staying with the Trouble: Making Kin in the Chthulucene* (2016), Donna J. Haraway refers to the contemporary era as a 'thick present' (p. 1). It is 'thick,' troubling, because we find ourselves—'all of us on Terra' (p. 1)—in a delicate and complex situation. In every region of the planet, we find situations of ecological damage. And in each of these situations, there are many different actors involved, who do not necessarily share a diagnosis, a position, a desire, or a proposal for action. As Haraway warns us, it is therefore difficult to arrive at a solution that solves the problem, repairs the damage, and simultaneously pleases all the actors involved. It would be too 'innocent' (p. 20) for us to arrive at a definitive solution. Instead, Haraway invites us to become capable of giving a responsible, and perhaps partial, response to the challenges and 'horrors' (p. 3) of the Anthropocene.

The lecture series *Thick Presents. The Arts of Living on a Damaged Planet* (named after Haraway's expression 'a thick present'), organized by the Instituto Valenciano de Arte Moderno (IVAM, Valencia, Spain) between 2020 and 2023, wanted to take up this invitation. The cycle took the ecological emergency of the Albufera Natural Park (Valencia) as a starting point, not with the innocent aim of turning it into a 'common cause' (Stengers, 2015) but to approach a concrete situation that emerges as a living echo of the knowledge put into circulation within the framework of the program. The lecture series relied on the knowledge and stories of the populations involved in the situation, which were in dialogue with a corpus of literary narratives and scientific texts that served as a critical tool.

In this text, we analyze *Thick Presents* program in order to think about the stories and practices that come into play in a situation of climatic emergency and ecosocial transition.

Around the arts of living on a damaged planet

Becoming capable of responding in a way that does not further hurt an already damaged earth is not an easy task. According to Isabelle Stengers (2015), we should start by stating that we are in fact 'ill-equipped' 'to produce the type of response

DOI: 10.4324/9781003489078-14

that, we feel, the situation requires of us' (p. 30). Why is that? Because what has disappeared from our everyday horizon, what progress and capitalist development have destroyed, is precisely what makes us capable, what enables us to be responsible, to think, and to resist, that is, the 'common causes' (pp. 87–88). It is not so much a question of whether each of us has innate abilities, or whether we are better or worse qualified to solve a given problem, whether we have a hierarchical (e.g. scientific) knowledge in relation to other knowledge. The answer, the success of an action, Stengers (2015, pp. 87–96) tells us, depends on our collectively re-engaging with the common cause that has been taken away from us. The common cause (an endangered wetland, river, or forest) has the power to bring us together and make us capable, to cause us to think, to feel, and to imagine a response that it would otherwise have rendered unthinkable.

The statement that we are badly equipped for responding to our present is thus revealed not as 'an observation of impotence, but rather of a point of departure' (p. 30). It is crucial to consider common causes as the ones which make us capable, because they take the individual out of the heart of debate and place the situation at the heart of the problem. Imagination and thinking are 'indissociable' from a cause and from 'a concrete, practical experience' (p. 132). Responses must therefore be formulated 'on a case by case, region by region basis,' and always in a way that gives a prominent place to 'the knowledges of interested people.' Thus, by building a common cause from a specific situation, we could invent a response to these catastrophic times (Stengers, 2015, pp. 128–132).

The lecture series *Thick Presents. The Arts of Living on a Damaged Planet* took the ecological emergency of the Albufera Natural Park (Valencia) as a starting point, not with the innocent aim of turning it into a common cause but to approach a concrete, real situation that could awaken the knowledge shared within the cycle, including the knowledge of the actors engaged in this specific situation. This kind of knowledge was considered 'the arts of living on a damaged planet' (Tsing, 2017), and it was as important as the artistic practices and literary texts that created a body of work in the context of *Thick Presents*. It is important to highlight this fact, which in fact serves as a foundation and a working methodology: there is no pre-established hierarchy between experiential, first-person, situated accounts of the agents who inhabit the Albufera, on the one hand, and the literary or artistic accounts of situations of damage and ecological transition that are convened in this context, on the other. This heterogeneous set of practices and narratives is intended to serve as a toolbox for thinking about imaginative actions in this situation of ecological collapse, not only the one that concerns the Albufera Natural Park, but also in other stances of environmental damage, following the proposal of Stengers (2015) and Andreas Weber (2016).

The Valencian Albufera Nature Reserve is one of the wetlands of highest environmental value in the Mediterranean Basin. It covers an area of 52,200 acres and is located just 10 km from the city of Valencia. The declaration of the Albufera as a Natural Park occurred in 1986. Since 1989, it has been recognized as a Wetland of International Importance, especially as a Waterfowl Habitat. Since 1990, it has also been included in the Special Protection Areas (Zepa in Spanish) and, in 2001, it has been classified as a Site of Community Importance. For these reasons, it is part of the Natura 2000 Network. Several areas within the park have also been declared Plant Micro-reserve and Wildlife Sanctuary.

Nevertheless, these figures have not led to appropriate measures to effectively protect the Park (its diverse forms of life), or to generate a response to the state of collapse to which it has been brought (GVA, 2022).

Currently, the conflict looming over the Albufera involves and concerns different actors, both humans and nonhumans. The government, farmers, fishermen, environmentalists, city dwellers, hunters, nearby towns and housing developments, industrial estates, migratory birds, and aquatic species are all agents that interact with each other with different—and sometimes opposing—interests and needs (Llorens i Dies, 2017; Piqueras, 2017; Fundació Assut, 2019; Dolç, 2021). For example, if rice farmers drain the lagoon beyond a certain limit, it is likely that birds will not find the conditions they need to rest and feed before undertaking the next stage of their migration. In this case, 'staying with the trouble' (Haraway, 2016) would mean not dismissing any of the viewpoints in this dispute but cobble together and feel, think, and imagine until reaching a noninnocent response that encourages regeneration and ensures that life-forms in the Albufera are being taken care of. In this example, the concern is not only for rice production and the continuity of agricultural livelihoods, but also for the lives of bird species, which are at risk if their migratory routes are not maintained (see Van Dooren, 2014). According to Haraway (2016), the task is 'to cultivate the capacity to respond to worldly urgencies with each other' in each 'thick present' (p. 14).

Program, format, and methodology of work: stories for a possible change

The *Thick Presents* program was designed for a working group of 25 registered attendees. All members committed to attending at least 80% of the sessions. It was structured around fortnightly activities (approximately) over an extended period of time (September–June). The working group meetings/activities were of diverse nature and included visits to the Albufera (guided by various agents), brief seminars, training sessions, reading groups, workshops, actions, walking routes, analysis of films, texts, and other artistic practices, etc. All the meetings of the working group were led by external guests or by the cycle curator.

The first edition of the lecture series (years 2020–2021) was as follows:

September 19, 2020: Guided tour of the Albufera Natural Park (I): Current status and problems of the Albufera National Park (ANP). Led by Javier Jiménez Romo (Biologist and environmental technician at the ANP). Current struggles to protect and take care of the ANP. Led by Víctor Navarro (Ecological Action-AGRÓ).

September 26, 2020: Guided tour of the Albufera Natural Park (II): Ornithological visit. About the birds in the ANP. Led by Mario Giménez (SEO/Birdlife).

October 10: Guided tour of the Albufera Natural Park (III): Living and dying in l'Albufera. Non-innocent relationships between companion species. Led by José Badía (Sueca Hunters Club).

October 24: Guided tour of the Albufera Natural Park (IV): The point of view of fishermen. Led by the fishing community of El Palmar.

November 7: Guided tour of the Albufera Natural Park (V): The point of view of rice farmers. Led by Vicent Moncholí (Farmer).

November 21: Internal working session: On the topic of the place and the perspective of non-human subjects in the Anthropocene era. Working session with a body of contemporary literary texts. Led by Miguel Ángel Martínez (cycle curator).

December 12: Seminar: Nature Writing: The literature of Nature. Led by Gabi Martínez (writer).

January 23: Seminar: Anthropocene, Capitalocene, Chthulucene: Donna Haraway's tentacular thinking. Led by Helen Torres (Sociologist and Donna Haraway's translator).

February 6: Seminar: Around the arts in the Chthulucene. Led by María Ptqk (Curator and researcher).

February 20: Workshop: About attention and imagination. Learning how to listen (including bird songs). Led by Carmen Pardo Salgado (Philosopher, educator, sound artist).

March 6: Internal working session: The bird scene. Led by María Jerez (Performer, playwright, educator).

April 10: Seminar: Forms and stories of mourning in cases of ecological damage (I). Artistic practices: Led by Mijo Miquel (Researcher, professor at UPV).

April 24: Seminar: Forms and stories of mourning in situations of ecological damage (II). Literary corpus: Led by Miguel Ángel Martínez (Cycle curator).

May 24: Seminar: *In Catastrophic Times*: *tiempos de catástrofes*: narratives, experiences, and working methods in cases of environmental conflict. Led by Miguel Ángel Martínez (cycle curator).

June 5: Internal working session: Group conclusions.

As seen in the program, the lecture series essentially consisted of two parts: the first block, in which the group conducted fieldwork in a series of tours to the Albufera Natural Park; and the second block that took place at the IVAM and was structured around an organized set of seminars, workshops, and internal training sessions (attended only by members of the working group).

The goal of the first block was to approach human and nonhuman ways of life in the Albufera Natural Park, in such a way that, by the end of the cycle, attendees would have a panoramic view of the ecological and social conflict looming over the territory. It was essential that this first objective be fulfilled before beginning the second block of the cycle, in which the session coordinators (the curator or the invited experts for each session) would provide the group with analysis tools that should sharpen their insights and viewpoints, through a comparative analysis between the circumstances of the ecosocial conflict in the Albufera and the analysis of other cases or more general features of the global ecological emergency crisis.

Therefore, the body of narratives used by the group throughout the cycle occupied a crucial position. This first part of this corpus were the stories told by the agents of the Natural Park—farmers, fishermen, ornithologists, hunters, technical staff, and ecologists—rooted in years of work experience in the Park, as well as in the memory of a life lived in the Albufera. The second part of this body of narratives were the literary texts and artistic practices that were the subject of analysis and work in the workshops, seminars, and group sessions of the second part of *Thick Presents*.

Regarding the first part of these narratives—the group's interviews, the stories shared during the meetings in the Natural Park, and some publications related to the territory—we would like to highlight the impact generated within the working group by the narrative approach taken by two different agents: on the one hand, hunters; on the other, biologists, technical staff, and ornithologists of the Park. We think that one of the main keys of *Thick Presents* program has been this impact, which emerged from substantial disagreement in the versions and discursive strategies of each group. This insight can be applied to the understanding of ecosocial conflicts, especially in relation to the role narratives play in these delicate situations.

The sequence of events was as follows: in the first session of the cycle, we were introduced to the current state of the Albufera by the environmental technician of the Park, Javier Jiménez Romo, and the biologist and founder of the organization Acció Ecologista-AGRÓ, Víctor Navarro. To accomplish this, they identified the factors that led the Natural Park to be on the brink of ecological collapse. A situation that, thankfully, has been gradually improving over the past two years mainly thanks to the contribution of water from the Júcar River received by the lagoon. This first session captured the interest of the group, as they were able to get an idea of the state of the Park, thanks to the mainly technical information provided by the guides. The second session followed the same path: Mario Giménez, delegate of the SEO/Birdlife organization in the Valencian community, informed, in purely scientific terms and without respite, the problems faced by the birds that inhabit or temporarily nest in the Park. As in the previous session, this information was of interest to the group, which, in any case, acknowledged the cascade of data, statistics, and scientific reports and undoubtedly felt the scarcely narrative nature of the ornithologist's speech.

The group's feelings regarding this first pair of sessions became clearer during the third visit to the Albufera, where we were received by José Badía, the president of the hunting club of Sueca. At the beginning of the session, the suspicion and distrust triggered by the presence of a hunter became evident. During the initial two sessions, the group have learnt that hunting was one of the problems faced by the birds of the Natural Park. Moreover, ecological awareness was prevalent in the group, as was to be expected; therefore, the practice of hunting was not highly regarded among its members. However, suspicion and distrust gradually turned into genuine sympathy, even though we suspected, or knew for certain, that some of the data provided by the hunter were incorrect or intentionally false (such as the effects of hunting, the number of specimens and species hunted). This transformation, which I found to be wonderful, was based, I would say, on two reasons: firstly, on the emotional connection that our interlocutor had with the territory.

It was evident—perhaps more for us than for him—that his hunting was born of and nurtured by that affect, because it was only one among all the other practices that shaped his way of life, which revolved almost entirely within the limits of the Natural Park. We were unable to discern if our previous interlocutors shared this fact. If we think about it, we could affirm that it was the case, that it cannot be otherwise. But we could not arrive at that conclusion from their interventions, it was not something obvious or easily perceptible to the group. If they felt that way, they did not convey it or share it. It remained hidden, imperceptible. This one, precisely, was the second reason why José Badía aroused such sympathy in the group: he openly displayed his affection for the territory. He did not say it explicitly, he did not name it, he did not even mention it, one could even say he was not aware of it. His feeling was orally crafted into his life stories. Despite the low expectations of the group, the hunter was a splendid storyteller. It was his stories what captivated the group. From this circumstance emerges a crucial hypothesis raised during the last session of the group that has to do with the consequences of the hunter's storytelling. On the one hand, it effectively served the narrator's goal: the group returned from that session with a more complex, less accusatory idea of hunting. On the other hand, and maybe despite the hunter's intention, the group increased its emotional commitment to the territory we were working on. Thanks to that life story, everybody felt that they knew the territory better, firsthand, which resulted in their ecological commitment and their defense of the territory, which undoubtedly included the birds of the Park. In the last session of the cycle, the group was aligned with the statement of Deborah Bird Rose, according to which individuals defend and take care of what they love (2011). Deborah Bird Rose referred, obviously, to climate emergency situations.

If the group reached this conclusion, it is probably thanks to the experience of having attended the other sessions of *Thick Presents*, where they became familiar with stories of ecological conflict and reflected upon the function and power of images and narratives in actions of care, defense, and repair of the planet. In this sense, some sessions are to be highlighted, such as the 6th, 7th, 12th, and 13th, in which we worked with a body of literary texts, artistic practices, and scientific publications from the fields of literary and artistic fields. In these sessions, coordinated by Gabi Martínez, Mijo Miquel, and myself, we worked, among other texts and short stories, with *H is for Hawk,* by Helen Macdonald (2014); *The Peregrine*, by J. A. Baker (2015); *The Birds*, by Daphne Du Maurier (1963); or *Flight Ways: Life and Loss at the Edge of Extinction*, by Tom Van Dooren (2014); *On the Animal Trail*, Baptiste Morizot (2021). In all the four sessions, there was an affective impact with the situations presented in the texts that became evident in varying degrees and in diverse forms. This was particularly noticeable during the sixth session, probably because the texts we worked on were about the lives of birds. The coincidence, in fact, was noticed and named by the group: once again, they had felt the emotion of engaged into a physical, embodied relationship with the protagonists of the stories we read, as it had happened with José Badía. The group acknowledged this fact when we commented on two texts by Vinciane Despret: *Living as a Bird* (2021a) and 'L'enquête des acouphènes ou les chanteuses silencieuses' (2021b). The first one is not only a critical study of the tradition of ornithological scientific research and an essay that underlines the importance of the insights of birdwatching lovers—who are amateur researchers—but also a text to honor bird song. The second one is a fictional narrative, which continues the work of Ursula K. Le Guin; a story in which a scientist is forced to incorporate artistic research methodologies into his research protocol thanks to spiders' actions, which are his object of study. I think that the participants felt touched by the 'performative nature' of the literary texts, it was the sense of the stories collected in the two texts what had the greatest impact on them. We also acquire this insight with the help of Jo Labanyi (2021), who urged us to pay attention not only to what texts mean, to the things they say but also to the ways they 'do' things, penetrate our bodies, and 'affect' us (p. 27).

In the last two sessions of the cycle, we reached a conclusion that I believe stemmed from the work of this previous series of meetings: despite the insistence of some members of the group, *Thick Presents* lectures were not, had not been, and did not want to be a process of citizen participation aimed at producing a specific political action. If anything, it had become a collectively conducted artistic research project based on the encounter with the Mediterranean territory of L'Albufera.

In any case, I do not see this fact, this conclusion, as a flaw, a lack, or an error in the group's working process. Actually, due to the working methodologies we put into practice, we believe that this assessment can be useful in some contexts and/or citizen participation processes aimed at a perfectly defined objective. We were not guided by a moral imperative to act in a tricky situation, nor by a control over the strategy to be followed to achieve a specific goal. Furthermore, constant presentiality was not a demand, as it is usually the case in activist organizations. In short, we lacked a productive common goal to be measured or quantified, that is, an objective that could later be used in the defense of the Albufera Natural Park. We were not so ambitious. We did not expect to get to that point. If anything, we believed in a sum of subjective, solitary transformations (Alemán, 2012), changes that took place at some point, thanks to the coexistence among group members, and that would eventually be updated and enacted through visible or invisible caring for the forms of life around us. A transformation that I would say was also of the order of the sensible (Rancière, 2004) and that had aroused from the fieldwork and the engagement with the stories we had read or heard aloud, alongside their protagonists.

Living again as a bird

I still remember very clearly the first time they surprised me. The sunlight, the beginning of a long afternoon, my wool jumper, and the silence. I remember very distinctly the silence before their appearance. I was sitting on the rooftop floor when they flew over me for a while, a brief lapse of time that I couldn't specify now. I remember a fleeting, intermittent green flash. But most of all, I remember hearing something that I couldn't tell if it was a conversation, a gathering, or an extraordinarily expressive singing.

What for me turned out to be an intense and singular memory is, in fact, a shared remembrance for the group. We all remember a moment when we stepped out onto the balcony or looked out of a window and heard a bird sing. My first encounter probably occurred during the third week of lockdown. The parakeets crossed the sky of my neighborhood in a line that took them directly to the old riverbed. Before that day, I had hardly noticed them. I had never even thought about why they were considered a 'plague.' Only when our way of life stopped and we remained silent for a while, could I (we) listen to the birds again.

The philosopher Vinciane Despret urges us not to forget this fact. But she also invites us to go a little further. If the only explanation we give is that it was the pause and the silence that allowed us to attend to that bird song, to that beauty, we will still be cornering the birds. 'As if the birds—warns Despret—didn't have their own point of view about what was happening to us. Did they maybe think that our silence meant that we had given them the floor?' (2020).

If we are to tell the story of lockdown, something undoubtedly necessary, we must do it differently. Vinciane Despret claims that we should include the story of the birds, the telling of how birds experienced lockdown (2020).

We do know, however, that birds changed their habits during the time humans were confined. Researchers who study recordings of bird songs have observed that, in many cities, birds 'sang more, and for longer periods,' during the COVID-19 crisis. 'They had to spend less energy fighting against noise and, surely, they could hear each other and understood themselves far better.' Therefore, we could say that we listened to them because both of us had 'liberated': we, humans, from our habits, from our hurry, from everything that prevents us from paying attention to what surrounds us; and the birds, from our 'anthropo-cacophony,' as Despret calls it (2020).

For the birds, it was a celebration. Moreover, it coincided with a crucial moment: the beginning of spring, which is when many of them prepare to inhabit a territory. As the Belgian philosopher reminds us, at this time of the year birds accompanied their songs with a display of colors and shapes, performing a visual and sonic festival. 'They unfold the magic of music, beauty, desire' (Despret, 2021a, p. 86).

This way, by taking all of these circumstances into account, we can tell a new story. A shared story. A story in which humans live in continuity with birds.

More than four decades ago, composer Bernie Krause began recording the soundscapes of Lincoln Meadows, in Sierra Nevada Mountains (USA). Some time later, in 1988, a logging company obtained permission to operate in that area, arguing that the method they would use—'selective logging'—would have no environmental impact. In this way, by cutting down a tree here and there, they said, nothing would change. If we look at the photographs taken before and after the operation, it seems they were not lying. Every tree, every branch, is in its place (2013).

However, 'our ears,' says Bernie Krause, 'tell us a very different story.' A year after the logging extraction, the composer returned to Lincoln Meadows. 'Using the same protocols and recording under the same conditions,' he gathered a good number of new recordings. Comparing the recordings before and after logging, we can realize that something indeed changes. In the latest recordings, we still hear the stream water, but the dawn chorus of birds has disappeared. Although we cannot see it in the pictures, the birds are no longer there: they had to leave their territory due to the effects of human action. 'I have returned to Lincoln Meadows 15 times in the last 25 years,' explains Krause, 'and I can tell you that the biophony, the density and the diversity of that biophony, has not yet returned to anything like it was before the operation' (2013).

Environmental sciences have typically tried to understand the world from what we see and have measured biodiversity indices by counting the number of species and the number of individuals within each species in a given area. However, 'by comparing the data that connect the density and diversity of what we hear,' we could obtain, according to Krause, 'much more precise results. Biophonies (animal sounds) and geophonies (earth sounds) are the signature voices of the natural world.' As we hear them, we can embrace 'the sense of a place, the true story of the world we live in' (2013).

During lockdown, when we stood out on the balcony to applaud or sing, we could embody that experience firsthand, that is, 'the power of a sung and sound space' (Despret, 2020).

As mammals, it is important for us 'to live our bodies in close contact with other bodies.' According to Despret, applause and music served precisely that function in a time of mandatory physical distance: they allowed us to touch the bodies of others from a distance, to be touched by them. Despret says that 'we, a group of human primates, invented a way to touch each other drawing inspiration from birds' (2020).

We can tell two different stories: the story of overcoming, of the fight against the virus, of how we managed to master nature, or 'a story that allows us not to forget the existence of birds' (Despret, 2020).

Program, format, and methodology of work: storytelling for a shared history

The first edition *Thick Presents. The arts of living on a damaged planet* started with an approach to the ecological damage situation in the Albufera, Valencia.

The Albufera represents a small dot on the map, a tiny area if considered on a planetary scale, or if viewed from the perspective of 'global warming' or 'climate change.' However, if we abandon this aerial, anthropocentric perspective and approach the trouble as earthly beings, as beings bound to the ground (Haraway, 1991, pp. 183–203), it reveals itself as a complex, 'thick' territory. This way, we are faced with a series of interrelated environmental problems that, at least apparently, do not have an easy solution—such as the level and quality of the lagoon water, the presence of 'invasive species,' the discussion about rice straw burning, the use of fertilizers in agricultural tasks, and the issue of hunting.

In light of this, the second edition of the lectures cycle proposed that we take more time and get a closer look at the territory. After we had listened to some of the people who live and work there and we had obtained an overview of the ecological conflict looming over the Albufera, we were able to approach a more specific situation, one of the issues that is part of a broader conflict, namely, the status of the birds that temporarily or permanently inhabit this wetland, declared a Natural Park 35 years ago.

Perhaps from that position, from this closeness, we might be able to continue the story of coexistence between humans and birds. Moving from the history of the Anthropocene to the promises of the Phonocene (Despret, 2021a).

The second edition of the lecture series had the following axes:

1 Continuity with previous work developed in the territory. Collaboration with the organization Acció Ecologista-AGRÒ
2 Taking a closer look. Getting down to earth. Listen. We moved from the axis *Albufera → Planet* of the first edition (where we approached the global climate emergency situation from the Albufera's specific ecological conflict), to the axis *Birds → Albufera → Planet*. Additionally, we moved from the trajectory *Anthropocene → Chthulucene* to the trajectory *Anthropocene → Chthulucene → Phonocene*
3 Getting closer also meant taking one step further toward practical work with the group. We not only continued to explore the territory and listen to its dwellers, humans, and nonhumans, but we also engaged in different activities with them. Practical work (fieldwork, hands-on work, etc.) connected with and enhanced the proposal of the cycle as a space for thinking
4 Continuity of two axes or entry points to the climate emergency situation: link with the territory and multispecies relationships
5 Paying attention to multispecies relationships as a point of departure, with the aim of acquiring a better understanding of nonhuman species and learning how to live with them. This learning would necessarily contribute to taking care of the space we share with these species. Basically, we tried to move away from what Eduardo Kohn (2017) calls 'cosmological autism' (p. 307)
6 Finally, to work once again with a body of life stories, literary texts and images, that allow us to bring affect into play (Labanyi, 2021).

As in the first edition, the second block of *Thick Presents* was addressed to a working group of 25 people (26 in this case), although we also included a series of activities open to the general public.

The working group was formalized through registration, and the attendees committed to attending at least 80% of the sessions. The group was ultimately composed of both individuals who had taken part in the previous edition (12), and newcomers (14). The cycle was also organized with weekly or biweekly activities over an extended period of time (January to July).

The activities and meetings were of diverse nature, led by external guests or by the cycle's curator. Throughout the cycle, we organized three birdwatching sessions, each in a different season of the year; we observed the migratory routes and corridors of birds; we recorded sounds and a 'dawn chorus' during a spring sunrise; and we learnt to track the different colonies that inhabit the park, in order to become familiar with their habits and learn to live with them with kindness.

The program for the second edition (year 2022) was as follows:

January 22. Guided tour of the Albufera Natural Park (I): Current status and problems of the Albufera National Park (ANP). Led by Javier Jiménez Romo

(Biologist and environmental technician at the ANP). Current struggles to protect and take care of the ANP. Led by Lucía Moreno (Ecological Action-AGRÓ).

February 5. Guided tour of the Albufera Natural Park (II): Birdwatching visit (I. Winter): Migration routes and disappearance of migratory corridors (Racó de l'Olla). With Diana Ferris (SEO-Birdlife).

February 12. Visit to the La Granja Wildlife Recovery Center (El Saler): Damage, care, and multispecies restoration. Conducted by the staff of the Wildlife Service of the Department of Agriculture, Rural Development, Climate Emergency, and Ecological Transition (GVA).

March 4. Performance + Seminar: Living Art, participatory rituals, and queer ecology in the face of ecocide (IVAM). By Graham Bell (2019) (artist and ornithologist).

March 11. Internal working session: *Flight Ways: Life and Loss at the Edge of Extinction.* Working session with a corpus of contemporary literary texts. Led by Miguel Ángel Martínez (cycle curator).

March 26. Sound landscape workshop and field recording (I) (ANP): Conducted by María del Castillo (researcher, sound artist).

April 2. Sound landscape workshop and field recording (II) (ANP): Conducted by Kamen Nedev (researcher, sound artist).

May 1 (dawn). Sound landscape workshop and field recording (III) (World Dawn Chorus Day) (PNA): Conducted by Kamen Nedev (researcher, sound artist) + Birdwatching session (Breakfast) (II. Spring) (PNA). With Lucía Moreno (Ecological Action AGRÓ).

May 5. Sound landscape workshop and field recording (IV) (IVAM): Conducted by Edu Comelles (researcher, sound artist).

May 6. Sound installation (IVAM): Assembly, installation, and inauguration of the work done in the sound landscape workshops. Conducted by Edu Comelles (researcher, sound artist) + working group.

May 21. Guided tour of the Albufera Natural Park (III): Bird tracking workshop (PNA). Conducted by Luisa Abenza (biologist, tracker, Cybertracker Spain trainer).

June 4. Guided tour of the Albufera Natural Park (IV): Scientific bird ringing session: migration routes and migratory corridors (Tancat de la Pipa). With Diana Ferris (SEO-Birdlife).

June 18. Guided tour of the Albufera Natural Park (V): Birdwatching Day (III. Summer) (Racó de l'Olla – Estany de Pujol). With Diana Ferris (SEO-Birdlife).

July 8. Performance: *Reclamo en el jardín* [Call in the garden] (IVAM), by María Jerez (performer, researcher, educator).

July 9. Internal working session: Group conclusions.

Note

1 This study has benefited from the support of the following institutions: Ministry of Science and Innovation, through the research project 'Crisis climática, salud mental y bienestar en el Antropoceno. Una aproximación desde la ontología histórica' (PID2021-124477OA-I00 funded by MCIN/AEI/10.13039/501100011033, by 'ERDF A way of making Europe').

References

Alemán, Jorge (2012). *Soledad: común. Políticas en Lacan*. Buenos Aires: Capital Intelectual.

Baker, John Alec (2015). *The Peregrine*. New York: Harper Collins.

Bell, Graham (2019). *Natural Hysteria (A Queer Response to Ecocide): An Exercise in Living Art, Participatory Rituals and Queer Ecology*. Valencia: La Errería.

Bird Rose, Deborah (2011). *Wild Dog Dreaming: Love and Extinction (Under the Sign of Nature Explorations in Ecocriticism)*. Charlottesville: University of Virginia Press.

Despret, Vinciane (2020). *Fonoceno*. Barcelona: La Bienal del Pensament, CCCB. https://www.cccb.org/es/multimedia/videos/las-voces-del-planeta/234586.

Despret, Vinciane (2021a). *Living as a Bird*. Medford, MA: Polity Press.

Despret, Vinciane. (2021b). Autobiographie d'un poulpe et autres récits d'anticipation. Arles: Actes Sud.

Dolç, Carles (2021). *Del Saler al Túria*. Valencia: Pruna Llibres / Alfons El Magnànim.

Du Maurier, Daphne (1963). *The Birds and Other Stories*. London: Penguin.

Fundació Assut (2019). *El cultiu de l'arròs a l'Albufera. Bones practiques*. Valencia: Fundació Assut.

GVA (2022). *Parc Natural de l'Albufera*. Generalitat Valencia. https://parquesnaturales.gva.es/es/web/pn-l-albufera.

Haraway, Donna (1991). *Simians, Cyborgs, and Women. The Reinvention of Nature*. New York: Routledge.

Haraway, Donna (2016). *Staying with the Trouble: Making Kin in the Chthulucene*. Durham / London: Duke University Press.

Kohn, Eduardo (2017). How dogs dream. Amazonian natures and the politics of transspecies engagement. *American Ethnologist*, 34(1) (February), 3–24. https://www.humanities.uci.edu/sites/default/files/document/kohn.pdf.

Krause, Bernie (2013). *The Voice of the Natural World*. TED Talk – TED Global. https://www.ted.com/talks/bernie_krause_the_voice_of_the_natural_world/transcript?language=es.

Labanyi, Jo (2021). Pensar lo material. *Kamchatka. Revista de análisis cultural*, 18, 15–31.

Llorens, Vicent y Dies, Bosco (2017): *Trenta anys, trenta mirades. Parc Natural de l'Albufera 1986-2016*. Valencia: Edicions 96 / Fundación Assut.

Macdonald, Helen (2014). *H Is for Hawk*. New York: Grove Press.

Morizot, Baptiste (2021). *On the Animal Trail*. Medford, MA: Polity Press.

Piqueras, Norberto (2017). *El Saler per al poble, ara!*. Valencia: Publicacions de la Universitat de València.

Rancière, Jacques (2004). The politics of literature. *SubStance*, 33(1), 10–24, Issue 103: Contemporary Thinker Jacques Rancière.

Stengers, Isabelle (2015). *In Catastrophic Times*. London: Open Humanities Press.

Tsing, Anna L. (2017). *Arts of Living on a Damaged Planet: Ghosts and Monsters of the Anthropocene*. Minnesota: University of Minnesota Press.

Van Dooren, Thom (2014). *Flight Ways: Life and Loss at the Edge of Extinction*. New York: Columbia University Press.

Weber, Andreas (2016). *The Biology of Wonder: Aliveness, Feeling and the Metamorphosis of Science Andreas Weber*. Gabriola: New Society Publishers.

14 Capital against nature

Accumulation by dispossession and externalization in the Mar Menor

David Soto Carrasco[1]

Introduction: a case of anoxia

On May 27, 2016, ANSE (Southeast Association of Naturalists) and WWF published a video denouncing the massive proliferation of algae that severely affected 80% of the Mar Menor coastal lagoon (Murcia, Spain). That recording was in itself a major milestone. The idyllic landscape of one of the most beautiful Mediterranean lagoons, listed as an internationally protected site (Ramsar, Specially Protected Areas of Mediterranean. Importance), European (Natura 2000 Network) and regional (partial Regional Nature Reserve), was becoming a gigantic "pea soup". The constant influx of nutrients, mostly nitrates from nearby intensive farming over decades, led to an "eutrophication" process (Aguilar et al., 2016; Ruiz-Fernández et al., 2020; Comité de Asesoramiento Científico del Mar Menor, 2017). The nitrates fertilized the algae and phytoplankton that proliferated unfettered, preventing any sunlight from penetrating the surface of the water and reaching the deeper parts of the lagoon, thus causing a drastic reduction in oxygen levels. Consequently, the water turned green due to the abundance of algae on the surface and the marine meadow, while fish died from suffocation. That "pea soup" made the headlines in the news of the printed and televised media, reporting a crisis that had been years in the making.

The Mar Menor, an age-old environment affected by a number of critical episodes throughout its history, was now facing an unprecedented crisis. From events such as deforestation on orders by King Felipe II to uncontrolled development in the 20th century, this natural ecosystem has borne witness to the complexities of interactions between human activity and the environment. Nevertheless, the current social crisis, characterized by deterioration of the water and loss of biodiversity, is a significant challenge that needs to be addressed immediately. The critical situation of the Mar Menor, denounced by ecologist associations for years, is the result of decades of institutional inaction and negligent policies that have permitted uncontrolled farming, mining, and industrial activities that are detrimental to the lagoon (ANSE & WWF-Spain, 2018). The lack of any effective regulations and the tolerance of illegal practices, such as effluent discharge, over-exploitation of aquifers, and the uncontrolled expansion of irrigation land, have further

DOI: 10.4324/9781003489078-15

led to environmental deterioration of the lagoon leading to today's collapse of the ecosystem (Martínez-Fernández, 2022).

In this context, this chapter propounds that the crisis of the Mar Menor is the result of complex interactions between human activity and the natural environment, particularly in the context of agro-extractivist practices. It further proposes examining such through interaction between the concepts of accumulation by dispossession and externalization, with the aim of critically understanding the underlying causes of this eco-social crisis, and to be able to put forward measures to mitigate it. In order to do so, I will mainly draw on two theoretical approaches, which I believe afford a complementary angle to each other: the "Four Cheaps" theory by Jason W. Moore, and Kohei Saito's externalization theory, in conjunction with the hypothesis of "capitalist realism" by Mark Fisher. The former provides an outlook on the economic dynamics that have led to the environmental deterioration of the territory, focusing on how the agro-extractivist model and the crisis of the Mar Menor are related. The second approach will provide an understanding of how externalization practices have exacerbated the environmental and social impacts on the ecosystem, discussing the technological, spatial, temporal, and psychological aspects of this problem. Finally, I will put forward the argument that the crisis of the Mar Menor is not only a paradigmatic example of environmental deterioration driven by the capitalist productive model in its agro-extractivist form, but it is also a case of ecocide through which it has become an environmental and social sacrifice zone where the contradictions inherent to neoliberal capitalism are blatantly laid before us.

Lagoon: a territory, a global emulsion

The process of ecological collapse of the Mar Menor has taken place in line with a multi-scale logic in which different social, economic, and political players are involved, operating at local, regional, state, and global level (Grove, 2019; Patel & Moore, 2017).[2] The Murcian lagoon and the different practices that affect its territory take place on different levels in a complex, multi-scale articulation of discourses, actions, and policies within the world-economy theory (Wallerstein, 1983, 2004, 2011). It comprises an emblematic landscape in the Region of Murcia, on which the intensive fabric of social and community relations linked to anthropologically valuable territory is built (Avilés Conesa, 2022). Not devoid of attacks in different shapes and forms, the Mar Menor underwent pressure from heavy metal mining, initially through the enormous amounts of effluent that were produced, followed by draining from the residual mining structures afterward (Martínez Barainca et al., 2023, pp. 55–61), only to suffer from the effects of urban development activity, including the associated tourism infrastructures (housing, artificial beaches and marinas) (Romero Díaz et al., 2017). Collapse of the Mar Menor in spring 2016, aggravated through the processes of 2019 and 2021, following two episodes of torrential rain that fed the lagoon with nutrients, thus increasing the mortality rate of fish, converted this marine environment in the paradigm of a critical zone exemplifying the contradictions of the agro-extractivist form of the productive economic model (Avilés Conesa, 2023; Martínez-Fernández, 2022; Martínez-Fernández & Esteve Selma, 2000). This contradiction, traditionally associated with the Global South, emerged in the very heart of developed Europe.

In the area adjacent to the Mar Menor known as Campo de Cartagena, traditional farming focusing on local and national consumption has gradually lost ground to give way to a more intensive, industrialized method of farming, mainly for the export of produce to the Global North (De Castro et al., 2017). Ferocious international competition arising through an increase in single-crop production and industrialization of farming in the Global South has increased pressure on the North's production patterns (Carrasco et al., 2012; Giraldo, 2019; Gómez-Barris, 2017; Perreault, 2013). This scenario drives territories such as the Campo de Cartagena and Mar Menor areas in the Region of Murcia to implement intensive, extractivist farming methods in line with the interests of the major multinational corporations in an international context, the dynamics of which have been described as the "new imperialism" (Crosby, 1986; Harvey, 2003; Huber, 2022). In the specific case of Campo de Cartagena, following completion of the Tagus-Segura Water Transfer System,[3] in the early 80s, the area of irrigated land multiplied tenfold, to reach 60,000 ha (Carreño, 2015; Martínez-Fernández, 2022). The Mar Menor region has unquestionably become very profitable for transnational agro-industrial investment since then, at the cost of freeing up assets at low or zero cost to appropriate them and make them even more profitable (De Castro et al., 2017; Pedreño, 1988). There are numerous global examples of this concept: oil wells, mega-mines, lithium, water, seas, deserts, etc. (Bebbington, 2009; Composto & Ouviña, 2009; Galafassi & Composto, 2013; Sacher 2015; Swyngedouw, 2005; Voyles, 2021). They exclusively operate on a profit-based or output-based logic, reducing the territories where they carry out their activity to abstraction sites, mirroring the concept traditionally described as "accumulation by dispossession" (Adrienne, 2008; Fine, 2006; Harvey, 2003, 2011; Little, 1999; Sparke, 2008).[4] It is a type of operation that inserts a regional area in the flow of global connection that demands an abundance of agricultural and natural products (Haesbaert, 2011; Taylor, 2015). In the case of the Mar Menor, as suggested earlier, this has taken place according to the specific logic of agro-extractivism. In our understanding, this concept leads to a certain degree of success, in that it has not only permitted proving the global logic of the process of accumulation over territories. It has also enabled a critical approach to humanity/nature dichotomization and has showcased the need for collusion among the social players who facilitate dispossession through establishing legislation that limits intervention by institutional powers in favor of liberalizing private players and pauperizing living conditions under a multifaceted logic.[5] Dispossession in the Mar Menor area took place at four moments in time: (i) the illegal expansion of land for irrigated crops; (ii) the use of technology and chemicals (fertilizers, pesticides, herbicides, etc.) that have degraded the soil and the water basin, leading to biological imbalance, a reduction in biodiversity and endangering the ecosystems; (iii) over-exploitation of groundwater through illegal wells (Hernández, 2024; Martínez-Fernández et al., 2013); and (iv) the use of the Mar Menor as a dumping ground for farming, industrial, mining, and urban development waste, with the subsequent impact this has on the natural balance and resilience of the lagoon (ANSE & WWF-España, 2018).

Although the concept of "accumulation by dispossession" significantly contributes to explaining what has happened in the Mar Menor: introducing the lagoon territory in the global market through the agro-extractivist logic of liberating assets at low cost, and also the consequences: economic growth, seizure of public or communal land and water, pollution of the ecosystems, etc., it could be insufficient to understand the devices that neoliberal capitalism deploys to seize communal assets, such as the seized land to grow crops in the territory, or to control nature. To further clarify, the most complex dynamics of appropriation are ultimately established through the humanity/nature dichotomy and the relationships of power, but also to analyze the latest contradictions of the accumulation process as such: the need to face the limits of production, of work and of nature need to be addressed (Angus, 2016; Velasco Santos, 2020). To do so, I will delve further into the "Four Cheaps" theory that are required in capitalist valuation, described by Jason W. Moore in his "World-ecology" that closely links the accumulation of capital with the production of nature (Moore, 2015, 2018a, 2018b, 2021, 2023a, 2023b). In short, it is not simply about endlessly proclaiming accumulation by dispossession but rather thinking about the humanity/nature dichotomy through a theory of accumulation, which at the same time is the consequence of the relations of power and joint production of nature in a dialectic relationship.[6] Under this line of logic, nature and humanity are historically and politically linked (Chakarbarty, 2009; Moore, 2014). That is how the dual internalness that characterizes Moore's thinking is established. History consists of humanity-in-nature and nature-in-humanity (San Román & Molinero-Gerbeau, 2023; Villegas, 2021). Human development and forces are therefore also natural, extra-human forces (Moore, 2015, p. 36).

The Four Cheaps comprise the unrecognized, appraised, or remunerated basis of capitalist production: workforce, raw materials, food, and energy (Moore, 2015, pp. 62–64). They are the essential elements for suitable development of the production system and represent the relations, spaces, and possibilities of valuable production work. Consequently, the absence of the "Four Cheaps" leads to a crisis caused by the system's need to have to face the costs of maintaining and expanding it. In this sense, capitalist remodeling of nature seeks to favor the ever-increasing accumulation of capital. On the specific subject of obtaining that "Cheap nature", it is achieved by expanding the frontiers of appropriation or dispossessing social groups or resources, which allows keeping costs low and sustaining capitalist accumulation (Moore, 2015, p. 74). In this light, the Mar Menor crisis is an example of how the "Four Cheaps" theory put forward by Moore can be interpreted. The ecological crisis is the direct result of a search for "Cheap Nature", where nature is perceived as an unlimited, exploitable resource. In the specific case of the Mar Menor, what was initially considered to be cheap: unlimited land for crops, water (brought in via the water transfer system) for irrigation, the produce that was grown, etc. were no longer cheap. The land was no longer cheap because of pollution and over-exploitation, water was no longer cheap because of climate change, etc., all of which was exacerbated by the increase in the price of energy.

Only the labor remained cheap by exploiting migrants from region of Maghreb and Sub-Saharan Africa (Pedreño et al., 2015). Therefore, the Mar Menor crisis should not only be seen as a process of accumulation by dispossession, but rather a precise compression of the surrounding area by human action and also intra-human relations (political) and between humans and nature, which are ultimately the relations that have led to the eco-social crisis of the biggest Mediterranean saltwater lagoon.

Therefore, the collapse of the territory around the Mar Menor could be included in a multi-scale context where different economic and political players at local, regional, state, and global level have taken part through strategies of accumulation by dispossession that have exacerbated the ecological and social crisis. This situation reflects how territory is inserted not only in the economy-world, but also in the ecology-world, and reveals the contradictions of the productive model, particularly in its agro-extractivist form (Navarro Trujillo & Linsalata, 2021). In other words, the Mar Menor crisis exemplifies how uncontrolled exploitation of nature leads to a crisis, a deep crisis that affects the ecosystems and the human communities that depend on them. Among others, local fishermen have lost their means of livelihood through low catches, homeowners have seen how their properties have devaluated, and the tourism industry has suffered serious losses because of pollution and deterioration of the area. This situation illustrates how the health of ecosystems and human well-being are interdependent, thereby, challenging the notion that nature can be exploited indefinitely without serious consequences, and above all the need to consider how aspects previously considered external are in fact internal contradictions of neoliberal capitalism.

Nitrates: nature against capital

What we have seen is that the collapse of the Mar Menor is inserted in a global context in which different players take part in strategic practices of accumulation by dispossession, which caused its ecological and social crisis. The Mar Menor crisis exemplifies how uncontrolled exploitation of nature leads to a deeper crisis that affects the ecosystems and the human communities that depend on them. This interdependence challenges the notion that nature can be indefinitely exploited without serious consequences and highlights the internal contradictions of capitalism. In this section, I shall discuss how Saito's theory of externalization (2023) provides a lens to help us understand how those contradictions play out in the context of the Mar Menor, and by drawing on Fisher's analysis (2009) on neoliberalism we will see a new externalization: the burden on mental health, also perceptible in the residents of the Mar Menor, and I shall conclude by presenting the crisis of the Mar Menor as an extreme case that highlights dissolution of the border in the transitional externalization of global costs.

Kohei Saito's (2017a, 2017b, 2020, 2022, 2023) Marxist interpretation on social and natural metabolism within the capitalist system delved further into how it undermines interaction between humans and nature right down to the ultimate consequences.[7] Capital, driven by the law of value, can only act unilaterally over nature, under ultimate compression in which nature and workers are seen as abstract work and appropriable raw materials. This logic conditions the metabolic human-nature relation, even compromising both of them. Further to Wallerstein's analysis, Saito (2022) upholds that the core of prosperous countries has been cheapening resources in peripheral areas and imposing the assumption of hidden charges for the beneficiaries of economic growth. This led Lessenich (2016) to decry developed nations as "externalization societies": societies which "externalize their problems". That is how climate change is perceived as a crisis of the capitalist system because, all told, it is a crisis of uncontrolled externalization. Furthermore, it allows us to perfectly exemplify the crisis of the Mar Menor as a moment showing accumulation by dispossession leading to the indiscriminate exploitation of the lagoon and its resources, but at the same time, transferring, i.e., externalizing, its contradictions to the territory and its communities, to the point of the ecosystem and the production system becoming unsustainable. In other words, if accumulation showed privatization of profits, externalization points to communitarization of the costs.

The first contradiction that Saito (2022) identified was the "technological transfer", which basically consists of the plundering of ecosystems. This first type of transfer consists of resorting to technological development in an attempt to overcome the climate emergency. It is not by chance that this is where Saito (2017a) draws on the criticism, cited by Marx regarding the chemist, Justus von Liebig's work, on the irrational management of farming, whereby the land is sacrificed at the altar for immediate profits, when calling for rational farming. The question of enormous requirements for fossil fuels to manufacture nitrates is further glossed over. In other words, for Saito (2022, pp. 36–39), today's farming cannot work without resorting to more and more quantities and varieties of chemical fertilizers and antibiotics, which are very expensive to manufacture and end up being released into the environment, thereby reducing the productive capacity of soil and, among other issues, altering the ecosystem.

The second type of transfer is "spatial transfer". This refers to the process whereby problems and negative external factors of capitalist production, such as pollution and depletion of natural resources, are geographically displaced to other regions or countries. This means that the areas where the goods and services for global consumption are produced suffer the environmental and social impacts of production, while the areas where they are consumed do not directly suffer those consequences.

The third type of transfer is "temporal". It refers to postponing or externalizing the negative impacts of production for the future, which means that the environmental and social costs build up over time. This leads to two consequences. On the one hand, future generations are not only at the mercy of unfavorable environmental conditions, but they are also economically jeopardized by the consequences of them, either through the impossibility of generating economic profits and working benefits, or because of the restrictions caused by the debts of the current costs of externalization.

I should also add a fourth type, namely, the psychological impact. This refers to the pressures and tensions of the capitalist system being individually internalized, which leads to people perceiving mental health issues, such as stress or anxiety, as if it were a personal problem rather than the result of broader socio-economic structures (Fisher, 2009, p. 19). This internalizing of stress, for example, contributes to disintegration and the feeling of hopelessness in contemporary society. Where the previous forms of externalization consisted of transference of the order of nature, time, and space, the latter consists of transference of the order of how an individual's subjectivity and psychic terrain are built. The contradictions and pressures of the capitalist system are transferred to the individual scope, making them appear to be personal problems instead of displays of a crisis of the economic and social system (Frazer-Carroll, 2023; Soto Carrasco, 2023).

The crisis of the Mar Menor exemplifies the concerns put forward by Saito (2022) on technological, spatial, and temporal transference and the impact thereof on ecosystems, in addition to the psychic transfer that has been outlined. The excessive use of chemical fertilizers in industrial farming producing for the global North has aggravated pollution in the territory, leading to an ecological and social crisis that highlights not only the importance of immediately implementing more sustainable farming methods but also the need to question the relationship between humans and nature. We need to bear in mind that the Region of Murcia is one of the most heavily polluted areas in the European Union caused by nitrates from farming. The maximum permitted amount of 50 mg/l of this kind of fertilizers was exceeded in this region. The Mar Menor was able to absorb around 3,300 metric tons of nitrates per year over several decades (Comité de Asesoramiento Científico del Mar Menor, 2017, p. 18; Ruiz-Fernández et al., 2020). The situation has hardly changed since the collapse in 2016. In 2022, 3,500 metric tons of nitrates continued to reach the lagoon (MITECO, 2023, p. 43). As mentioned at the beginning, the nitrates asphyxiated the Mar Menor, but it is now rebelling against the very model of capitalism. The situation does not only depict an eco-social crisis, but it can also be construed under the concept of ecocide, understanding such to be the deliberate or negligent destruction of a vital ecosystem constituting a crime against nature and humanity (Baquedano Jer, 2019; Broswimmer, 2003; Neira et al., 2019; Zierler, 2011). At the same time, it shows that the transference of environmental loads is reaching their limit, and that the intensifying contradictions of the current economic model are starting to be seen on a large scale in the global North (Parker, 2013).

Some examples of this are the overloading of nitrates and phosphates in the water of the Mar Menor, leading to minimum catches through traditional fishing techniques (Salas, 2013; August 11). The deterioration that the Mar Menor has undergone in recent years has also had an effect on the local economy. Deterioration of the lagoon led to a reduction in gross family earnings of around 100 million Euros between 2015 and 2019. More specifically, the populations on the shores of the Mar Menor felt the brunt of it during that period with a direct reduction of 7% of their gross family earnings, and indirectly of 11% compared to the unaffected areas, thus leading to a total reduction in gross family earnings of 18% (Maté-Sánchez-Val & Aparicio-Serrano, 2023). The effects of the crisis were seen not only in tourism, catering, or leisure in the area, but also in other industries such as farming and construction, the latter being linked to the loss of real estate values. Deterioration of the lagoon has also meant that tourists decide to choose other coastal regions, thus increasing business failures (Aparicio et al., 2024; Maté-Sánchez-Val & Aparicio-Serrano, 2022). Furthermore, we should not forget that living in areas affected by pollution and uncontrolled exploitation negatively affects physical and mental health (Berry et al., 2010; Fritze et al., 2008). Deterioration of the Mar Menor in recent years has affected young people in the area in particular, as they are more susceptible to the effects of pollution and extreme climate changes. Four out of ten young people living in this area have noticed a worsening of their quality of life due to health problems associated with the conditions of the lagoon (García, 2021, September 19).

In short, the crisis of the Mar Menor provides some tangible examples of the concepts of technological, spatial, temporal, and psychological transference. Firstly, excessive dependence on chemical fertilizers in intensive farming in the region is a way of technological transference where the search for higher production yields has led to the pollution of the aquatic ecosystem. Overloading the water of the Mar Menor with nitrates through decades of unsustainable farming practices shows how environmental problems are transferred over time and space, affecting future generations and the surrounding areas. Moreover, the resulting devastation of the ecosystem, the effects on human health, and the local economy reflect psychological externalization. These specific examples underscore the urgent need to find a solution to the crisis of the Mar Menor and the need for a change to the production model through a comprehensive approach that acknowledges this externalization and deals with it in order to guarantee the sustainability and health of the ecosystem and the human communities that depend on it.

Final considerations: Mar Menor, sacrifice zone

The collapse of the Mar Menor clearly illustrates the complex interactions between human activity and the environment, and the serious consequences of agro-extractivist practices within the context of neoliberal capitalism. This case provides us with a panoramic outlook on how indiscriminate exploitation of natural resources, in line with the capitalist logic of accumulation by dispossession, within a framework lacking effective regulation, triggers an ecological and social crisis of alarming proportions. Through Jason W. Moore's outlook on the Four Cheaps required for capitalist valuation, I have analyzed how the Mar Menor has historically been a cheap nature within the context of expanding international demand for cheap food. Furthermore, by exploring the collapse of the Mar Menor through Saito's theory of externalization, we have seen how the negative effects of capitalist production are displaced into the future, space and individual psyche, triggering a chain of consequences that impact the ecosystem and the human communities that depend on it, and who perceive it as such (Guaita-García et al., 2021). Overloading the Mar Menor with nitrates, resulting from decades of unsustainable farming practices and policies, is a clear paradigm of how those transferences affect the health of the ecosystem, the traditional economy in the region, and the quality of life of the people who live there. These items clearly show an economic practice involving an ultimate political and institutional compression: capitalism of profits and communism of costs and waste.

The plundering of resources, externalization of costs, and transference of environmental and social impacts to nature, communities, and people reveal the deep underlying contradictions in our current relationship with the environment. This is how the crisis of the Mar Menor reflects the radicalization of environmental injustice under the concept of "sacrifice zones" and how it is expanding to more and more important areas (Juskus, 2023, Klein, 2014; Latour, 2014; Svampa, 2013; Svampa & Viale, 2014). In other words, when talking about the Mar Menor, we are not only talking about an area that has undergone environmental deterioration, but rather one that has undergone a significant deterioration of the well-being of local communities. The situation is the result of an aggressive economic, political, and social model that prioritizes immediate private profit without considering the repercussions that compromise the long-term life and future of the Mar Menor.

Notes

1 This publication is part of the R&D&i projects "Climate crisis, mental health and wellbeing in the Anthropocene. An approach from historical ontology (ANT-mentalhealth)" (PID2021-124477OA-I00) and "Study and criticism of Italian Theory" (PID2021-123958N), funded by MCIN/AEI/10.13039/501100011033/ and ERDF "A way of making Europe".

2 In multi-scale dynamics, "the global" and "the local" are presented as a process in which alliances between transnational businesses and national, provincial and local institutions, who promote a certain development model are crystallized; and however the local communities – self-organized residents, NGO, cultural groups, ecological organizations, provincial representatives of political parties, who question that model and the lifestyles that are imposed through it (Svampa & Viale, 2014).

3 The Tagus-Segura Water Transfer System is an aqueduct in Spain carrying water from the River Tagus to the River Segura water basin. It started operating in 1979. It led to a radical change to coastal farming in the southeast of Spain, enabling industrial transformation to intensive farming in the territory, leading to huge social and labor mutations in the social structure of the towns and villages in the Campo de Cartagena and Mar Menor areas (Pedreño, 1988).

4 David Harvey (2003, 2011) coined the term "accumulation by dispossession" to explain the contradictions of social production and reproduction inherent to neoliberalism. It is based on the Marxist concept of "primitive accumulation", which analyzes the divide between the worker and the proprietor of his working conditions. Harvey identified different processes such as commoditization and privatization of the land, converting property rights in exclusive rights, transforming the workforce into merchandise as a way of accumulation by dispossession.

5 Law 3/1987, of April 23, on the Protection and Harmonization of Uses of the Mar Menor was repealed by Law 1/2001, of April 24 on Land in the Region of Murcia by the "Partido Popular" political party, which fuelled urban development and the expansion of unregulated farming. The lack of control and eutrophication of the Mar Menor in 2016 led to the approval of a new law two years later to regulate arable and livestock farming activity in the areas near to the lagoon. In 2018, the Regional Government of Murcia passed the Urgent Mar Menor Measures Act, establishing three zones vulnerable to nitrates with different levels of obligations. This legislation also prohibited the use of highly soluble, potentially polluting fertilizers, such as ammonia nitrate, calcium nitrate,

and urea (Salazar-Ortuño, 2020; Vicente & Salazar, 2022). Today, the far right political party Vox is lobbying for new legislation in the Murcia Regional Parliament to reduce the level of protection established in the 2018 law in favor of the agro-extractivist industry. Regarding the constant breach of the Law, and more specifically the regulations on nitrates pollution in the Mar Menor territory, the following can be consulted: Giménez Casalduero (2023).

6 According to Moore (2015), the continuity of primary accumulation reconfigures reproduction associations, both human and environmental, in order to enable the extended, continuous flow of work, food, energy, and raw materials at a low cost within the merchandise system. This constant need to reorganize alters the order of relations of power, and therefore political logic is always disputed.

7 There is a significant debate between Foster (2000, 2013, 2024), Moore (2015, 2021, 2023b), and Saito (2022, 2023) about social metabolism and the value theory that involves different political strategies. Nevertheless, it is here that we limit ourselves to tracking the items that they agree on in their different theories to understand the eco-social crisis of the Mar Menor. For further information on the full debate, I recommend consulting the following, among others: Huber and Phillips (2024), Villegas (2021).

References

Adrienne, R. (2008). Privatizing social reproduction: The primitive accumulation of water in an era of neoliberalism. *Antipode*, *40*, 535–560.

Aguilar Escribano, J., Giménez-Casalduero, F., Mas Hernández, J., & Ramos-Esplá, A. A. (2016). *Evaluación del estado y composición de la comunidad fitoplanctónica de las aguas del Mar Menor*. Universidad de Alicante.

Angus, I. (2016). *Facing the Anthropocene*. Monthly Review Press.

ANSE & WWF-España (2018). *La burbuja del regadío: el caso del Mar Menor. Evolución de los regadíos en el entorno del Mar Menor*. Campo de Cartagena (1977–2017), Anse.

Aparicio, G., Camacho, M., & Maté-Sánchez-Val, M. (2024). Quantifying the Impact: Are Coastal Areas Impoverished by Marine Pollution? Retrieved from SSRN: https://ssrn.com/abstract=4635941, https://doi.org/10.2139/ssrn.4635941.

Avilés Conesa, Á. D. (2022). Ontologías duales en el campo de Cartagena: El Mar Menor en disputa. *Revista Andaluza de Antropología*, *22*, 89–103.

Avilés Conesa, Á. D. (2023). El Mar Menor zona crítica: agro-extractivismo, agricultura y sociedad. In J. Turpin Saorín (Ed.), *Antropología en devenir político: encuentros trans-disciplinares* (pp. 69–86). Dado Ediciones.

Baquedano Jer, S. (2019). Ecocide or environmental self-destruction. *Environmental Ethics*, *41*(3), 237–247.

Bebbington, A. (2009). The new extraction: Rewriting political ecology in the Andes. *NACLA Report on the Americas*, *42*(5): 12–20. https://doi.org/10.1080/10714839.2009.11722221.

Berry, H. L., Bowen, K., & Kjellstrom, T. (2010). Climate change and mental health: A causal pathways framework. *International Journal of Public Health*, *55*, 123–132. https://doi.org/10.1007/s00038-009-0112-0.

Broswimmer, F. (2003). *Ecocide. A Short History of the Mass Extinction of Species*. Pluto Press.

Carrasco, A. E., Sánchez, N. E., & Tamagno, L. E. (2012). *Modelo agrícola e impacto socio-ambiental en la Argentina: monocultivo y agronegocios*. Asociación de Universidades Grupo Montevideo y Universidad Nacional de La Plata.

Carreño Fructuoso, M. F. (2015). *Seguimiento de los cambios de usos y su influencia en las comunidades y hábitats naturales en la cuenca del Mar Menor, 1988-2009, con el uso de SIG y teledetección* (Doctoral dissertation, Universidad de Murcia).

Chakarbarty, D. (2009). The climate of history: Four theses. *Critical inquiry*, *35*(2), 197–222.

Comité de Asesoramiento Científico del Mar Menor (2017). Informe integral sobre el estado ecológico del Mar Menor. Retrieved from: https://canalmarmenor.carm.es/wp-content/uploads/2020/07/Informe-Integral-sobre-el-estado-ecol%C3%B3gico-del-Mar-Menor.pdf (Accessed 24 Apr 2024).

Composto, C., & Ouviña, H. (2009). Acumulación por despojo y nuevos cercamientos: Mercalitización de los bienes comunes y antagonismos renovados en América Latina. In *V Jornadas de jóvenes investigadores*. Instituto de Investigaciones Gino Germani, Facultad de Ciencias Sociales, Universidad de Buenos Aires. Retrieved from: https://www.aacademica.org/000-089/71.pdf (Accessed 24 Apr 2024).

Crosby, A. W. (1986). *Ecological Imperialism*. Cambridge University Press.

De Castro, C., Gadea, E., Pedreño, A., & Ramírez, A. (2017). Coaliciones sociales y políticas en el desarrollo del sector agroexportador: las frutas murcianas en las redes globales de producción agroalimentaria. *Mundo Agrario*, *18*(37), e043. https://doi.org/10.24215/15155994e043.

Fine, B. (2006) Debating the "new" imperialism. *Historical Materialism*, *14*, 133–156.

Fisher, M. (2009). *Realism Capitalism: Is There No Alternative*. Zero Books.

Foster, J. B. (2000). *Marx's Ecology: Materialism and Nature*. NYU press.

Foster, J. B. (2013). Marx and the rift in the universal metabolism of nature. *Monthly Review*, *65*(7), 1–19.

Foster, J. B. (2024). Extractivism in the anthropocene. *Monthly Review*, *25*(3). Retrieved from: https://monthlyreview.org/2024/04/01/extractivism-in-the-anthropocene/ (Accessed 24 Apr 2024).

Frazer-Carroll, M. (2023). *Mad World: The Politics of Mental Health*. Pluto Press.

Fritze, J. G., Blashki, G. A., Burke, S., & Wiseman, J. (2008). Hope, despair and transformation: Climate change and the promotion of mental health and wellbeing. *International Journal of Mental Health Systems*, *2*, 1–10. https://doi.org/10.1186/1752-4458-2-13.

Galafassi, G., & Composto, C. (2013). Acumulación por despojo y nuevos cercamientos: el caso de la minería a gran escala en la Patagonia argentina. *Cuadernos del CENDES*, *30*(83), 75–103.

García, A. (2021, September 19). El deterioro del Mar Menor pasa factura a la salud del 37 % de los niños de la zona. *La Opinión de Murcia*. Retrieved from: https://www.laopiniondemurcia.es/comunidad/2021/09/19/deterioro-mar-menor-pasa-factura-57431078.html (Accessed 21 Apr 2024).

Giménez Casalduero, M. (2023). El Mar Menor y la contaminación por nitratos: una situación tóxica sin solución jurídica efectiva. In L. Sánchez Gallardo (Ed.), *Mirando a los ríos desde el mar. Viejos y nuevos debates para una transición hídrica justa* (pp. 29–35). Editum.

Giraldo, O. F. (2019). Political Ecology of Agriculture. Springer.

Gómez-Barris, M. (2017). *The Extractive Zone: Social Ecologies and Decolonial Perspectives*. Duke University Press.

Grove, J. V. (2019). *Savage Ecology: War and Geopolitics at the End of the World*. Duke University Press.

Guaita-García, N., Martínez-Fernández, J., Barrera-Causil, C. J., Esteve-Selma, M. Á., & Fitz, H. C. (2021). Local perceptions regarding a social-ecological system of the mediterranean coast: The Mar Menor (Región de Murcia, Spain). *Environment, Development and Sustainability*, *23*(2), 2882-2909. https://doi.org/10.1007/s10668-020-00697-y.

Haesbaert, R. (2011). *El mito de la desterritorialización. Del fin de los territorios a la multiterritorialidad.* Siglo XXI Editores.

Harvey, D. (2003). *The New Imperialism.* Oxford University Press.

Harvey, D. (2011). *A Brief History of Neoliberalism.* Oxford University Press.

Hernández, K. (2024). *La contaminación por nitratos y su impacto en el medio ambiente y el agua de consumo humano. Estudio de los datos de contaminación de las aguas superficiales, subterráneas y de consumo humano de 2022.* Ecologistas en Acción.

Huber, M. T. (2022). *Climate Change as Class War: Building Socialism on a Warming Planet.* Verso Books.

Huber, M. T., & Phillips, L. (2024). El comunismo decrecentista de Kohei Saito: «Empezar de cero». *Jacobin Latin America.* Retrieved from https://jacobinlat.com/author/matt-huber-y-leigh-phillips/.

Juskus, R. (2023). Sacrifice zones: A genealogy and analysis of an environmental justice concept. *Environmental Humanities, 15*(1), 3–24.

Klein, N. (2014). *This Changes Everything: Capitalism vs Climate Change.* Simon & Schuster.

Latour, B. (2014). Some advantages of the notion of "critical zone" for geopolitics. *Procedia Earth and Planetary Science, 10,* 3–6. https://doi.org/10.1016/j.proeps.2014.08.002.

Lessenich, S. (2016). *Neben uns die Sintflut. Die Externalisierungsgesellschaft und ihr Preis.* Carl Hanser Verlag.

Little, P. E. (1999). Environments and environmentalists in anthropological research: Facing a new millennium. *Annual Review in Anthropology, 28,* 253–284.

Martínez Barainca, M., Baños Páez, P., Muñoz Maluenda, C., & Peñas Castejón, J. M. (2023). El impacto de la minería en las aguas de la Sierra Minera de Cartagena-La Unión. Implicaciones en las zonas de borde costero adyacentes. In L. Sánchez Gallardo (Ed.), *Mirando a los ríos desde el mar. Viejos y nuevos debates para una transición hídrica justa* (pp. 55–61). Editum.

Martínez-Fernández, J. (2022). *El Mar Menor. Falacias y realidades.* Fundación Nueva Cultura del Agua.

Martínez-Fernández, J., & Esteve Selma, M. A. (2000). Estimación de la entrada de nutrientes de origen agrícola en el Mar Menor mediante un modelo dinámico. *Mediterránea, 17,* 5–30.

Martínez-Fernández, J., Fitz, C., Esteve Selma, M. A., Guaita, N., & Martínez-López, J. (2013). Modelización del efecto de los cambios de uso del suelo sobre los flujos de nutrientes en cuencas agrícolas costeras: El caso del Mar Menor (Sudeste de España). *Ecosistemas, 22*(3), 84–94.

Mate-Sánchez-Val, M., & Aparicio-Serrano, G. (2022). Seawater deterioration and the tourist beta convergence process: A geospatial big data analysis of the Spanish Mediterranean coast. *Current Issues in Tourism, 25*(18) 1–17. https://www.tandfonline.com/doi/full/10.1080/13683500.2021.2021156.

Mate-Sánchez-Val, M., & Aparicio-Serrano, G. (2023). The impact of marine pollution on the probability of business failure: A case study of the Mar Menor lagoon. *Journal of Environmental Management, 332,* 117381. https://doi.org/10.1016/j.jenvman.2023.117381.

MITECO (2023). Marco de actuaciones prioritarias para la recuperación del Mar Menor: informe de Avances. Diciembre 2023.

Moore, J. W. (2014). The end of cheap nature, or, How I learned to stop worrying about 'the' environment and love the crisis of capitalism. In C. Suter, & C. Chase-Dunn (Eds.), *Structures of the World Political Economy and the Future of Global Conflict and Cooperation* (pp. 285–314). LIT.

Moore, J. W. (2015). *Capitalism in the Web of Life: Ecology and the Accumulation of Capital*. Verso.
Moore, J. W. (2018a). How to read capitalism in the web of life: Towards a world-historical materialism in the web of life. *Journal of World-Systems Research*, *21*(1), 153–168. https://doi.org/10.5195/jwsr.2022.1127.
Moore, J. W. (2018b). The capitalocene part II: Accumulation by appropriation and the centrality of unpaid work/energy. *The Journal of Peasant Studies*, *45*(2), 237–279. https://doi.org/10.1080/03066150.2016.1272587.
Moore, J. W. (2021). Empire, class and the origins of planetary crisis: The transition debate in the web of life. *Esboços: histórias em contextos globais*, *28*(49), 740–763. https://doi.org/10.5007/2175-7976.2021.e83493.
Moore, J. W. (2023a). On capitalogenic climate crisis: Unthinking man, nature and the anthropocene, and why it matters for planetary justice. *Real-World Economics Review*, *105*, 23–34.
Moore, J. W. (2023b). Our capitalogenic world: Climate crises, class politics & the civilizing project. *Studia Poetica*, *11*, 97–122.
Navarro Trujillo, M. L., & Linsalata, L. (2021). Capitaloceno, luchas por lo común y disputas por otros términos de interdependencia en el tejido de la vida. Reflexiones desde América Latina. *Relaciones Internacionales*, *46*, 81–98.
Neira, H., Russo, L. I., & Álvarez Subiabre, B. (2019). Ecocidio. *Revista de filosofía*, *76*, 127–148.
Parker, G. (2013). *Global Crisis: War, Climate Change, & Catastrophe in the Seventeenth Century*. Yale University Press.
Patel, R., & Moore, J. W. (2017). *A History of the World in Seven Cheap Things: A Guide to Capitalism, Nature, and the Future of the Planet*. University of California Press.
Pedreño Cánovas, A. (1988). *Del jornalero agrícola al obrero de las factorías vegetales. Estrategias familiares y nomadismo laboral en la sociedad murciana*. Universidad de Murcia.
Pedreño Cánovas, A., De Castro Pericacho, C., & Gadea Montesinos, M. E. (2015). Asalariados inmigrantes en enclaves de agricultura intensiva: crisis del sur de Europa y sostenibilidad social. *Papeles de relaciones ecosociales y cambio global*, *131*, 49–59.
Perreault, T. (2013). Dispossession by accumulation? Mining, water and the nature of enclosure on the Bolivian Altiplano. *Antipode*, *45*, 1050–1069. https://doi.org/10.1111/anti.12005.
Romero Díaz, A., Caballero Pedraza, A., & Pérez Morales, A. (2017). Expansión urbana y turismo en la Comarca del Campo de Cartagena-Mar Menor (Murcia). Impacto en el sellado del suelo. *Cuadernos de turismo*, *39*, 521–546.
Ruiz-Fernández, J. M., León, V. M., Marín-Guirao, L., Giménez-Casalduero, F., Alvárez-Rogel, J., Esteve Selma, M. A., Gómez Cerezo, R., Robledano Aymerich, F., González Barberá, G., & Martínez Fernández, J. (2020). Informe de síntesis sobre el estado actual del Mar Menor y sus causas en relación a los contenidos de nutrientes. Boletín Contencioso Administrativo, *2*(3), 4–12.
Sacher, W. (2015). Megaminería y desposesión en el Sur: un análisis comparativo. *Iconos. Revista de Ciencias Sociales*, *51*, 99–116.
Saito, K. (2017a). *Karl Marx's Ecosocialism: Capital, Nature, and the Unfinished Critique of Political Economy*. NYU Press.
Saito, K. (2017b). Marx in the anthropocene: Value, metabolic rift, and the non-cartesian dualism. *Zeitschrift für kritische Sozialtheorie und Philosophie*, *4*(1–2), 276–295. https://doi.org/10.1515/zksp-2017-0013.

Saito, K. (2020). Marx's theory of metabolism in the age of global ecological crisis. *Historical Materialism*, *28*(2), 3–24. https://doi.org/10.1163/1569206X-20202802.

Saito, K. (2022). *El capital en la era del Antropoceno*. Penguin Random House.

Saito, K. (2023). *Marx in the Anthropocene: Towards the Idea of Degrowth Communism*. Cambridge University Press.

Salas, A. (2013, August 11). La pesca del Mar Menor cae un 90% y la Cofradía teme tener que echar el cierre al no haber ingresos. *La Verdad*. Retrieved from https://www.laverdad.es/murcia/pesca-mar-menor-cae-cofradia-teme-tener-20230811005850-nt.html (Accessed 21 Apr 2024).

Salazar-Ortuño, E. (2020). La protección del Mar Menor bajo la normativa de la Administración. *Boletín Comisión Contencioso-Administrativo Juezas y Jueces para la Democracia*, *2*, 13–15.

San Román, Á., & Molinero-Gerbeau, Y. (2023). Anthropocene, capitalocene or westernocene? On the ideological foundations of the current climate crisis. *Capitalism Nature Socialism*, *34*(4), 39–57. https://doi.org/10.1080/10455752.2023.2189131.

Soto Carrasco, D. (2023). Políticas del terror: subjetividad neoliberal y populismo autoritario. In J. Turpín Saorín (Ed.), *Antropología en devenir político* (pp. 243–264). Dado Ediciones.

Sparke, M. (2008). Political geography-political geographies of globalization III: Resistance. *Progress in Human* Geography, *32*, 423–440.

Svampa, M. (2013). Consenso de los Commodities y lenguajes de valoración en América Latina. *Nueva Sociedad*, *24*. Retrieved from: https://nuso.org/articulo/consenso-de-los-commodities-y-lenguajesde-valoracion-en-america-latina/ (Accessed 21 Apr 2024).

Svampa, M., & Viale, E. (2014). *Maldesarrollo: La Argentina del extractivismo y del despojo*. Katz Editores.

Swyngedouw, E. (2005). Dispossessing H2O: The contested terrain of water privatization. *Capitalism, Nature, Socialism*, *16*(1), 81–98.

Taylor, M. (2015). *The Political Ecology of Climate Change Adaptation: Livelihoods, Agrarian Change and the Conflicts of Development*. Routledge.

Velasco Santos, P. (2020). El ciclo hidropolítico en Tlahuapan, Puebla: reflexiones en el capitaloceno. *Collectivus. Revista de Ciencias Sociales*, *7*(2), 51–72. https://doi.org/10.15648/Collectivus.vol7num2.2020.2673.

Vicente Giménez, T., & Salazar Ortuño, E. (2022). Los derechos de la naturaleza y la ciudadanía: el caso del Mar Menor. *Revista Murciana de Antropología*, *29*, 15–26. https://doi.org/10.6018/rmu.524761.

Villegas, A. (2021). El ecomarxismo entre el Antropoceno y el Capitaloceno: rupturas metabólicas, capital fósil y régimen ecológico. *Colombia Internacional*, *108*, 15–38.

Voyles, T. B. (2021). *The Settler Sea: California's Salton Sea and the Consequences of Colonialism*. University of Nebraska Press.

Wallerstein, I. (1983). *Historical Capitalism*. Verso.

Wallerstein, I. (2004). *World-Systems Analysis: An Introduction*. Duke University.

Wallerstein, I. (2011). *The Modern World-System I: Capitalist Agriculture and the Origins of the European World-Economy in the Sixteenth Century* (Vol. 1). University of California Press.

Zierler, D. (2011). *The Invention of Ecocide: Agent Orange, Vietnam, and the Scientists Who Changed the Way We Think about the Environment*. University of Georgia Press.

15 Minor emotions

How to live in a damaged environment

Juan Manuel Zaragoza[1]

Introduction

> Concrete. Cement. For sale. To rent. Falling apart. Rotting. Bitten cables. Defeated porches. Half-built houses. Dust and containers. Vacant lots with debris. Broken architectures. Worm-eaten old building materials. Ostentatious buildings. Beach bars and kiosks with a CLOSED sign. Shop fronts with white-washed windows. An open tobacconist's. An English pub with four drunks.
>
> (Méndez, 2023, p. 15)

This is the start of *Lodo [Mud]*. Begoña Méndez's book, although it was not the first published book about the "catastrophe" of the Mar Menor,[2] it was by far the most acclaimed by those involved in the defence of the lagoon. This is rather interesting, because the description of the Mar Menor by the author never goes far from what is expressed in the first paragraph. "And I sunk ankle-deep in mud. The texture was disgusting and my skin was itching" (Méndez, 2023, p. 30). In the post-apocalyptical place described by Méndez, what deserved to be saved? Or beyond that, what made the Mar Menor a place deserving of the emotional commitment shown by its residents? Méndez replies early on to the latter question: memory.

> I can imagine their comments: *La Manga of my childhood, La Manga of my summers and my first love; La Manga of the seahorses floating with me in the water*.
>
> (Méndez, 2023, p. 25)

But this, as Méndez points out, does not lead anywhere. Looking back at the past only causes melancholy and impotence: "From the identity node of men and their environment sprouts a regionalist pompous environmentalism that is useless" (Méndez, 2023, p. 25). Again, that memory-based link seems senseless. Fruitless, "sick of humanity", Méndez says of La Manga (2023, p. 31). What drives someone to fight for it? Further still, what future can be imagined for it by those involved in its defence?

DOI: 10.4324/9781003489078-16

Leaving aside Méndez's book, in this chapter I would like to understand the type of emotional links established between a concrete ecosystem such as the Mar Menor and the humans who inhabit it, and how those emotions can contribute to their mobilisation in its defence. I consciously use the word ecosystem because, as I'll defend later on, the case of the Mar Menor is not about a loss of landscape but a loss of the ecosystem values of the lagoon, which explains the particular experience of loss in this context.

The Mar Menor: from catastrophe... to hope?

The Mar Menor is the largest saltwater lagoon in Europe and an ecosystem with unique characteristics which has served, for instance, as a refuge for the noble pen shell clam (Belando-Torrentes et al., 2015), an endangered Mediterranean mollusc since the arrival, via the Suez Canal, of a parasite which kills it. However, this parasite cannot overcome what until recently was the main feature of the Mar Menor and which, as told in the first chapter of this book, explained its delicate balance: its high salt levels.[3]

Over the last 50 years, the Mar Menor, a paradisiacal location, has undergone the pressure derived from, firstly, the excessive urbanisation of its coastal areas for tourism purposes and secondly, the growth of a powerful intensive agricultural industry in its catchment area. This growth has caused pollution by nitrates in the reservoir adjacent to the Mar Menor and an increased entry of freshwater in the lagoon. Its effect, as Robledano et al. reminded us in Chapter 2, was detected for years, particularly with the increasing disappearance of the wetlands adjacent to the Mar Menor as well as the birds which inhabited it. All this pressure was first translated in a proliferation of algae in 2016, in what was known as *green soup* and, particularly in the high death toll of sea life on October 12, 2019, when over four and a half tonnes of fish, crab, prawn, seahorses, etc. had to be removed from the coast. And this does not include the death toll of small molluscs which later studies have estimated as a loss of 956,642 million specimens.

As a result of this catastrophe, a citizen movement was launched to demand legal personality for the lagoon. This movement brought before the Spanish parliament a popular initiative which collected, in the midst of the COVID pandemic, 639,000 signatures in the whole country, that is 28% more than required by law. This movement was led by a so-called promoter group and numerous activists formed its "hard core"; they travelled around the country to secure signatures (Salazar Ortuño & Vicente Giménez, 2022; Vicente Giménez & Salazar Ortuño, 2022). Following the success of this legal initiative, which was passed by the Spanish parliament in September 2022, the Mar Menor became in fact the first European ecosystem with legal personality. A part of this group, together with members of the Banderas Negras [Black Flag] movement, started an association, AMARME [Embracing the Mar Menor].

The pilot study I will discuss in the following pages was carried out with 35 members from this and other associations. The majority were women aged 60 and over, and many of them had retired. The men, far fewer in number, had a similar profile. The aim of this study was to understand the type of emotions that had driven them to get mobilised in such a committed manner. However, before describing the session and analysing the participants' responses, I would like to point out to some important aspects to understand the emotional impact of the Mar Menor crisis.

The intrusion of Gaia and the founding of a people

> [Margarita:] That morning I was told there were dead fish in the Villananitos beach and I went there on my bicycle, with a friend. When we arrived, my heart sank. All those fish out of the water. Gasping. Asking for air. I burst out crying.[4]

The daily routine of many people, who until then was pleasantly spent, as is common in the Spanish east coast in autumn, enjoying mild temperatures and days which are long enough for people to enjoy the evenings, was interrupted that October morning. The Spanish National Day, a Thursday, was marked by the presence of thousands of dying fish and shellfish in the seashore. There had not been any warning, not even the day before, that a catastrophe of these proportions could take place. Nothing in the previous days had deviated from an experience repeated multiple times, no dissonance, no warning.[5] Only the categorical intrusion of death. An intrusion which would become a question and eventually action.

How can the intrusion of a dying sea turn into a question? This is something that would come up later, when we asked the group, but it is not whimsical at all. Isabel Stengers was the first author to talk about the power of nature, of Gaia, to pose questions from its absolute indifference:

> It is up to us to create a manner of responding, for ourselves but also for the innumerable living species that we are dragging into the catastrophe, and, despite this "us" only existing virtually, as summoned by the response to be given [...] What we have to create a response to is *the intrusion of Gaia.*
>
> (Stengers, 2015, pp. 41–42)

Stengers insists on refusing to be trapped by a powerful name (*Gaia*) so that we think about what naming it creates. Gaia, she claims, is blind. She is indifferent to our reaction, that is why our reaction will never be "to Gaia", but to what provoked its intrusion and its consequences. "It is a matter here of thinking *intrusion, not belonging*" (Stengers, 2015, p. 44). The intrusion of Gaia, its unexpected appearance in our lives, provokes rupture, uprooting. That is why we cannot think of ourselves as belonging to a world that is no more (because now it *is* Gaia), but as an intrusion which can produce a new us, which so far only exists virtually. Which must be created to inhabit the intrusion. A new "us" for a new "world".

It is possible to read about what happened in the Mar Menor that terrible morning in October 2019 as that *intrusion* (local, perhaps) of Gaia. An intrusion in a concrete and localised place. In a complex ecosystem, such as a coastal lagoon interwoven in a wide network of social, economic, political webs... And if the matter was about posing the question not thinking about belonging but about intrusion, who of us will think of intrusion of the Mar Menor (that little Gaia) in our lives? "All of us", replied Margarita, echoing Teresa Vicente's words, "all of us who love the Mar Menor".

"639,826 signatures", Margarita repeats passionately. "I include all of them, each and every one of them, down to the last 26. Because all of them have contributed to what we have achieved". The signature collection process, which made possible the success of the popular legislative initiative, had the unexpected effect of the founding a "people", an *us*. A strange people, for sure, whose existence is, perhaps, lesser (Lapoujade, 2021), because it is virtual, because it is *possible*. But a people who, at the same time, exists because it has effects – as strange as the people itself. People who made possible the passing of a law which is unique in Europe, but who are incapable of changing a regional government which claims to be the main responsible for the estate of the lagoon. People who were only founded through their action, their response to the intrusion. Be it a collection of signatures or an Embrace to the Mar Menor which, barely a couple of years later would become an Embrace to Water, as it unifies struggles, demands and experiences, riverbeds, reservoirs and wetlands throughout Spain (Rivas, 2023). But it is incapable of making its presence felt at the ballot box.

Fragmented scenarios of an environmental damage experience

Sierra Minera extends to the south of the Mar Menor. Having been exploited since 2500 BC until the beginning of the 1990s, the impact of the mining activity on the Sierra and its environment – from Portman to the Mar Menor – is obvious.[6] Furthermore, and unsurprisingly, with regard to the value of its landscape. However, contrary to what could be expected, the experience is not necessarily negative; rather the opposite: even this damaged landscape can elicit a profound aesthetic experience, however aware we are of its environmental implications (Alcaraz León, 2022). Thus, our experience of environmental damaged is not inevitably influenced by a clear awareness of the damage, leading to *grief*, or, in some extreme cases, to *solastalgia* ("We feel solastalgia", Margarita says several times). The environmental damage is not a monolithic experience as we find many nuances, some unexpected, such as appreciating the beauty of a sunrise reflected on the stagnant water of a pond polluted by mining residues.

Facing Sierra Minera, degrading because of its exploitation, there is the Mar Menor. The lagoon unfolds before us so that, on a clear day, when the easterly wind blows away the morning mist, we can see its full extent. From our vantage point, we can see its islands (Barón, Perdiguera, as well as Sujeto and Ciervo). In the southern end, closest to us, we see the abandoned salt flats of Marchamalo, now in the process of restoration. The furthest away, the northern end, where the coast turns eastwards looking towards the Mediterranean sea, we see in the distance the brightness of the San Pedro salt flats, still operating. The sun reflects on the calm sea. The buildings in La Manga, suspended between two seas, seem to float and

Isla Grosa is silhouetted against the horizon. To the right, the Cabo de Palos lighthouse stands, watchful. It is as if we were in front of a landscape which has not been altered in millennia. Who could suspect that this place has been the scenario of an environmental disaster of such magnitude? What are the trails and traces which allow us to infer that a terrible ecocide has taken place in this unaltered environment?

There is nothing in the landscape to warn us that here, just a few years ago, an environmental disaster of such magnitude took place.[7] One of the largest in Europe. Let's go for a moment to Portman, on the other side of the Sierra. There, between 1957 and 1990, the French multinational SSM Peñarroya – España dumped 60 million tonnes of mineral chemical waste in the sea, through the Roberto washing site pipelines (Ortín, 2020). As a result of this environmental attack, the bay of Portmán disappeared completely, clogged with highly polluting mining residues which forever changed the landscape and the lives of the inhabitants of the small coastal town. The abandoned Portman nautical club is the only proof, for outsiders, that the sea was once there. The berths where the boats once moor now rust in front of a polluted land surface, where nothing or nobody walks.

But nothing like this happens in the Mar Menor. One of the particularities of this terrible catastrophe, this ecocide, is that the landscape has not undergone any transformation. This is why it is so difficult to explain to outsiders that the current Mar Menor is the result of an environmental catastrophe and because of it, it is an endangered ecosystem, terribly damaged, which needs to be *restored*. But what needs to be restored? When to the naked eye, *nothing has changed*. And this is again the greatest peculiarity of this case: that there is no immediate evidence of the catastrophe, that all evidence, all proof, is always mediated. By science or by memory – however sterile this may appear to Begoña Méndez.

This feature is not inconsequential because evidence of environmental damage tends to be linked to certain effects on the emotional well-being of the implied populations. Thus, for instance, Hwong et al. made a distinction between direct and indirect effects of climate change on mental health. The former include after-effects related to trauma or post-traumatic stress, normally associated with large-scale catastrophes such as hurricanes or flooding. Indirect effects are however more insidious because, firstly, their causes do not tend to be so obvious (rising temperature, pollen count increase, etc.) nor are their effects, which are linked to feelings of defencelessness or worry (Hwong et al., 2022, p. e281; Zaragoza, 2024).

The concept of solastalgia – with which, as seen above, the participants are familiar– also originated with the radical changes in the composition of the landscape which led to a feeling of pain, in connection to the loss of a place that we consider our own:

> Solastalgia [...] is the pain experienced when there is recognition that the place where one resides and that one loves is under immediate assault (physical desolation). It is manifest in an attack on one's sense of place, in the erosion of the sense of belonging (identity) to a particular place and a feeling of

> distress (psychological desolation) about its transformation. It is an intense desire for the place where one is a resident to be maintained in a state that continues to give comfort or solace.
>
> (Albrecht, 2005, p. 45)

In Albrecht's case, the definition was associated with the struggles in his region (New South Wales, Australia) against the expansion of open-cut mining exploitations: a particularly damaging practise which causes major changes in the composition of the landscape. Thus the emphasis on the physical desolation of the inhabited place in the definition of 'solastalgia' (Albrecht et al., 2007, p. 96; Albrecht, 2011, p. 50; 2019, p. 27). This does not mean that we cannot use the term in our case, of course, but it makes it much less *obvious*, because the direct experience of the environmental catastrophe in the Mar Menor can only happen in certain moments of ecosocial crisis, what we have termed in other texts as *ecosocial drama*.[8] In other words, in dramatic situations, such as the *anoxic* episodes of October 2019 and August 2021. The rest of the time, our experience of environmental damage will always be mediated: by the contrast between the current state of the lagoon and our memories of a distant past and, in many cases, idealised, either by a series of scientific data requiring a considerable mediation task to be able to mobilise spirits and emotions. This is why it is important to note, as we pointed out in the introduction, that this is not a case of loss of landscape but something more complex and which was already described masterfully in the first chapter of this book: the loss of an ecosystem balance.

Obviously, I am not stating that the perception of a landscape is immediate. To do such a thing would be to ignore all the literature on landscapes written in the last few years in disciplines as diverse as anthropology, history, visual culture studies and human geography (Tilley, 1994; Schama, 1996; Mitchell, 2002a, b). What I am saying is that part of what constitutes "a landscape" is its capacity to be perceived *as if it were* immediate, that is, as if we perceived absolutely *that which is really there*:

> The overall impression of the scene, though, transcended all particulars. The outlines and shadowed depths of the cliffs seemed archetypal: in all the transience of things, somehow this moment revealed the true and original textures of the landscape. It was as if I'd been granted for a minute an untarnished perception of things. Or as if my look had been washed clean, disinfected, so that it was precisely no longer a look, and all things could be unfiltered, unaffected by it.
>
> (Wylie, 2009, pp. 275–276)

It is precisely this capacity of the landscape that prevents the perception of environmental damage in the Mar Menor, unless is strongly mediated by science (continuing monitoring of oxygen levels, of the murkiness of the waters, etc.), or the constant comparison with the past (Was the water so murky *before*? Was there so much mud *before*? Was the sensation I felt *before* the same one I am feeling *now*?).

If the landscape had been lost, if the Mar Menor had stopped providing its beautiful sunsets, it would have been easier to understand and to make people understand what happened. And also, let us not fool ourselves, *keeping up* the constant claim for the need of justice and reparation.

But this has not been the case and therefore turning to science and memory are the only elements capable of making us understand what happened. That is why Begoña Méndez was mistaken. From the memory (deeply idealised, we agree) of the Mar Menor of the past does not "sprout a regionalist and pompous environmentalism that is useless" (Méndez, 2023, p. 25), but a citizen movement which has forever changed not only the Spanish legal system but, as pointed out by Teresa Vicente and Eduardo Salazar in Chapter 9, the whole logic behind western law with the introduction of a new paradigm: environmental justice. In the following sections, I will try to shed some light on the emotional factors which made it possible to achieve this.

A day which is impossible to describe, my soul emptied

The fifth session of the course took place in Llano del Beal, a town linked to mining and the workers' struggles at the end of the 19th century and beginning of the 20th century, which shaped its character. In many aspects, the town seems to be trapped in that moment in history. Particularly the houses. They are the typical dwellings you can find everywhere in the Campo de Cartagena area. One storey, with a pitched roof and a large patio at the back. The majority have a narrow facade. Barely 5 m long, with two openings: one for the door, one for the window. Others are as long as 8 or 10 m long and have a second window. This type of building, based on the traditional architecture of the area, was very common up to the 1920s and 1930s. Rehabilitation of buildings, after the Spanish Civil War, particularly from the 1950s onwards, would change both the building techniques and the materials, and most of all, its aesthetics.

The Casa del Pueblo, owned by UGT (one of the main national unions, linked to the socialist party PSOE), opened in 1916. Designed by Víctor Beltrí (renown Modernist architect of that period), it was built in barely three years. It is a large 34 x 16 m column-free hall with a pitched roof. It is simply designed with three low arched windows in the façade, two three-bay and one five-bay windows, in the *Seccession* style. The colourful floors, decorated with geometrical motifs, are striking. So is the coffered ceiling, with exposed timber means, as a result of recent restoring. At the back, there is a stage.

In front of the stage, on a long table, the materials needed for the session were laid out. A3 sheets of paper, scissors and all several types of pens, pencils, glues, magazines... At that point, all the course participants know each other and new relationships and trust-based connections have been created. However, the division between activists and non-activists is still obvious. One can see, for instance, in how close each group is, something which is difficult to break up.[9]

The aim of the session, split into two parts, was to explore the emotions that link us to the Mar Menor. In the first, more theoretical part, I talked to them, with the help of a brief presentation, about how we connect to nature through several practises. For instance, *shinrin-yoku* or forest baths, a practise imported from Japan which is about walking *consciously* around the forest; or meditation practise in nature (Rohan, 2023). We also talked about eco-anxiety and solastalgia, two expressions which had already come up in other sessions. But what was really important happened in the second part.

"We're going to make a zine". "But … what's a zine?" None of the participants knew the word. A zine is nothing else than a small-circulation print-work, made non-professionally by an individual or a small group of people. The format allows for a considerable stylistic freedom, and mixing text and image (photographs, illustrations, collages) are common. Sizes also vary, ranging from a paper fold that can fit in a pocket to something similar to a book (Biel & Brent, 2022). But the key thing about this format, what is really radical about a zine, is that it allows us to express, quickly and simply, by using our imagination and letting us be led by artistic intuition, ideas and feelings. Verbalising them into a more complex discourse would be much more difficult; thus for the intervention I am describing here, it is a very efficient tool (Zaragoza, 2024). It is a very good way to shape and share a subjective experience which often we have not been capable of elaborating until then, either because of lack of time to reflect on it, or because it contains traumatic elements (Zaragoza, 2021).

I suggested the following. We would make a zine following a classic folding form. This would allow us to have eight "pages" in an A6 booklet, big enough to be able to work with comfortably but small enough to keep it in a pocket. To make it, we would use any material on the table we had prepared and we could use any technique we wished, but it had to contain images and words. Furthermore, the zine had to follow the following structure: the cover page and the back cover would be left blank, so that the participants could work on it at home. On the first couple of pages, they had to express their connections with the Mar Menor – when did their relationship with the lagoon start, what were the memories of that time, what experiences left an imprint, etc. For the following two pages, I told them we would work on painful emotions because that was the time to face the damage the Mar Menor had endured, to relive the eco-social drama. At that point, I showed a photograph on the projector of the dying fish on the beach in October 2019. For this, the questions were more direct: Where were you? What did you do? How did you feel? For the following two pages, I asked them about their activism: Why were they driven to action? What was the process like? What did they find? etc. These are some of the results.

Links

In the first "task", we focussed on exploring our personal connection with the lagoon. The aim was to try and identify our first experience of the Mar Menor, the oldest possible memory we could find. For some of the participants, this was quite easy, as they had recent memories. In this profile of participants, mostly were retirees who bought a property in the Mar Menor as a second home, so remembering that "first meeting" was not difficult. In contraposition, others spoke of a longer

Links with the Mar Menor

1

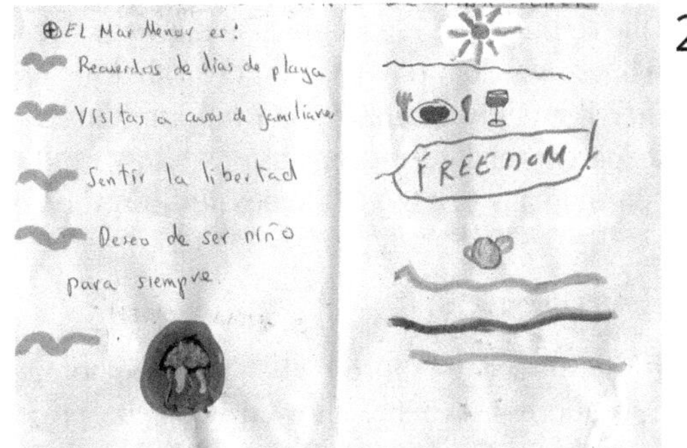

2

Ecosocial Drama

3

4

Figure 15.1 Reproduction of zines made in the workshop. Corresponding to "Links with the Mar Menor" and "Ecosocial Drama".

relationship with the Mar Menor, frequently going back to their childhood, but they would find it impossible to identify a particular experience. When this happened, they talked about unconnected memories, often linked to abstract concepts and ideas. If we pay attention to the first image in Figure 15.1, it reads:

> The first thing that comes to mind when I think about my connection with the Mar Menor are my grandparents and my cousins. Those bright and fun mornings at the Barnuevo beach in Santiago de la Ribera.

This is not the only connection with childhood. For instance, the second image reads:

> The Mar Menor is: Memories of beach days; Visiting relatives; Feeling free; Wishing to be a child forever.

That mixture between memory and desire, that sensation of freedom linked to the memories of a summertime with multiple free unsupervised time by adults multiplied resulted in that wish of "Wishing to be a child forever" at the end of the list. On the right, written in English, is the word "Freedom".

These two examples make it clear that the emotional connection with the Mar Menor is mainly established during childhood, in connection with the figure of grandparents and peers (cousins or siblings) and a series of practises which allowed the participants to expand the possibilities for experience.

Ecosocial drama

The third image in Figure 15.1 shows the photograph that the session's participants had to address. When I showed it on the projector, and despite having warned them, their comments of disgust multiplied. Despite the passage of time, the image continued to stir many feelings, some of them very profound. Again, there was an agreement in the words used: rage, anger, impotence. The text of image 4 reads:

> Anger, anger, anger and even more anger, followed by impotence, frustration and resentment against everybody who caused this: power, avarice, corruption, selfishness and contempt towards nature - the free beauty that only gives, does not demand anything in return. Loneliness and grief.

The rest of comments shared a similar tone and, in many cases, also feelings. In the group of people linked to the popular initiative, via Banderas Negras or the promoter group, not only the same feelings, expressions or words are shared (for instance, "impotence" appears in all their comments) but also a certain way of understanding nature as a generous, maternal entity that is mistreated by their abusers unjustifiably.

Activism

Curiously, this was the aspect least alluded to. Possibly because it was brought up last and there was not enough time to cover it. In any case, it is possible to draw some conclusions. Firstly, the continuity between the ecosocial drama and the activism process must be pointed out, even though it was expected. The feelings mentioned above such as rage and impotence were explicitly mentioned, but now, in the knowledge that they were shared, they helped as a driver to action: "Union, collaboration and participation. Emotion drove us to action", one of the zines reads. "Shared feelings", reads another.

Furthermore, significantly, a new word, a new emotion appears – hope. The hope that it is possible to revert the damage. This also tends to be linked to the success of the popular initiative to obtain legal personality of the Mar Menor, and a paradigmatic change with regard to our relationship with nature, which is expressed very clearly albeit in a not very elaborate manner: a text reads "Seeing nature from something to be exploited to something alive that we can save by getting together", accompanied by the drawing of a seahorse and the text "Mar Menor popular initiative".

Mediations/deep feelings

Before moving on to the conclusions, I would like to dwell on two aspects which, although weren't suggested directly by me, appeared to be very obvious. The first is related to something already pointed out: the need for mediation not only to understand what is going on in the Mar Menor, but also to establish those links we explored in the workshop. Although some participants mentioned sailing (with several references to Escuela de Pieter, one of the first sailing schools in the Mar Menor), and swimming, by far the preferred activity to talk about the connections with the lagoon was snorkelling:

> I could snorkel and spent hours going in and out to the surface, observing everything surrounding me, which was wonderful. I've got some fantastic turquoise memories.

This was not only experienced during childhood but it was present in the accounts of those who arrived at the Mar Menor later; vividly and fundamentally:

> Bright and shallow waters, tangled blue crabs, green meadows dancing in the bottom, and a little seahorse dancing slowly before my eyes. That was my first seahorse in the Mar Menor.

The images, as seen in Figure 15.2 (1 and 2), also point to the importance of snorkelling experience in establishing these connections.

The second aspect that came up during the workshop, that I would like to mention, was related to the expression by some of the female participants, of a deep connection with the Mar Menor, to the point of complete identification with it.

For example, image 3 in Figure 15.2 is a drawing and a poem by one of the group "leaders". The single-line drawing shows a female figure on the left, representing the woman, and in her heart, two lines converge to give shape to the Mar Menor – outside her but directly and intimately connected to her. The poem says the following:

> You are part / of my person / of my life / of my existence.
> This is why I fight / for you / so that you are / a part of all of us women.

Some of the participants wanted to share their ideas with the group. When she tried to explain what this drawing meant, she burst out crying and did not manage to finish her presentation.

The second zine I would like to talk about is shown in images 4 and 5. The first thing that drew my attention was, of course, the technique used, which I will discuss later. Before that, it is important to point out that this female participant did not follow the instructions I gave at the beginning of the workshop. In particular, she skipped the third section, devoted to activism and focused on the second, about

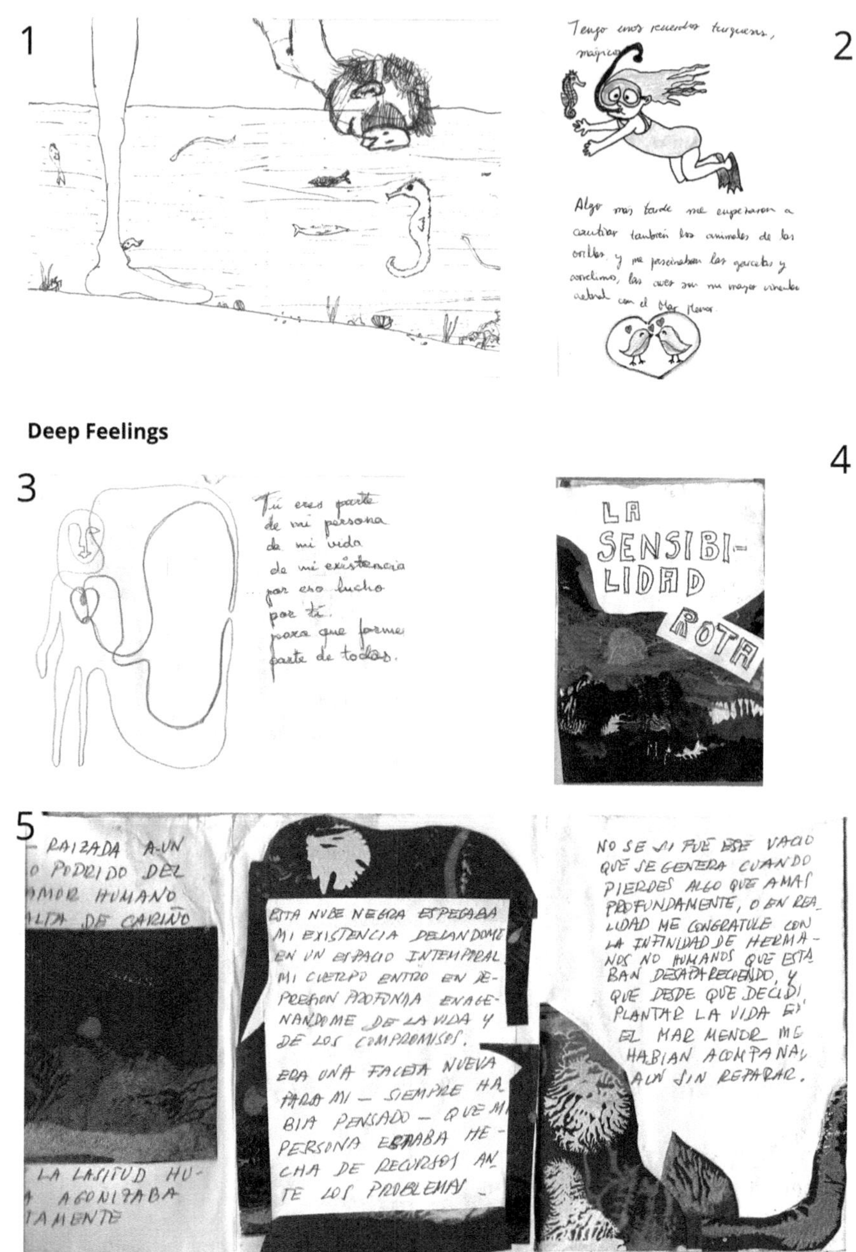

Figure 15.2 Reproduction of zines made in the workshop. Corresponding to"Mediations" and "Deeps Feelings".

the dramatic moment of the fish on the seashore gasping for oxygen. She devoted five pages of her zine to the ecocide. The text reads as follows:

> A day which is impossible to describe, my soul emptied … My breathing afflicted me and the magic of living evaporated like dried out petals when there is no rain.
>
> My life in tatters, I wander but I couldn't express my pain, and a cloud darkened within me.
>
> I was petrified, I couldn't get out of my trance.
>
> My feet froze and my hands were petrified.
>
> I was rooted to a rotten soil because of the human lack of love. Its lack of affection faced with human laxity agonised slowly.
>
> This dark cloud thickened my existence, leaving me in a timeless space.
>
> My body entered a profound depression which alienated me from life and commitments.
>
> This was a new facet for me. I always thought that my person was made [sic] with resources to tackle problems.
>
> I don't know whether that is the emptiness that occurs when you lose something that you love deeply, or that deep down I was happy for the countless non-human brothers who were disappearing and who, since I had decided to settle my life in the Mar Menor, had been my companions even though I had not noticed [them].
>
> The whole that had taken place in my existence had started to crack.
>
> That whole ended up as a sigh so broken that I didn't know how to react.
>
> I was left on my own on my paper boat, waiting for somebody to rescue me.

The tone of the text is complemented by the technique she used. The artist made a sort of collage with black cardboard paper on which, with a speckled or sponging technique, she created an effect of textures and degradation effect in forms that look like the bottom of the sea. The darkness of the composition seems to reflect the darkness which, according to its author, "thickened her existence".

As in the case above, what we see in this zine is the identification between the author and the lagoon, in this case by presenting the parallelism between the Mar Menor dying and the "deep depression" the author entered. In the text, supported by the illustration, the connection is obvious and we cannot stop imagining that the black colour upon which the sea forms are outlined is the colour of the author's "inside", which resulted in a "cloud darkened within me". The title of the zine was "Broken sensitivity". When sharing her experience, she also burst out crying.

Conclusions

As I mentioned in the introduction, the aim of this chapter was simply to try to understand the connections that are established with an ecosystem – The Mar Menor, in this case – and how these can mobilise their inhabitants to defend it. I have particularly paid attention to the emotional links because, although there are other links which impact activism – financial, for instance – when the respondents were asked what had driven them to activism, they unfailingly responded that their primary motives were emotional. This brought me, on the one hand, to be interested in when and how those links were established and, on the other, to understand how the ecosocial drama that took place in the Mar Menor made these and other relationships, normally assumed implicitly, move to the foreground. In this context, I would like to highlight two elements which I feel are fundamental.

The first aspect is the role played by memory in this process. As we saw in the introduction, authors such as Begoña Méndez point to memory turning into a cul-de-sac when it becomes a conservative nostalgia which would end up as "pompous, regionalist environmentalist that is useless" (Méndez, 2023, p. 25). This transformation of memory into useless conservative nostalgia is not however instinctive – this is where Méndez is mistaken. In the case of the Mar Menor, as we have seen, memory is not sterile but is fundamental in a double process: firstly, as mediation that allows us to confirm the delicate environmental state of the lagoon in contrast with the past; secondly, and because of the above, as a driver for activism in the defence of the Mar Menor, with the aim to *recover it*. As one of the zines reads: "Our Mar Menor will return to its full life because my soul feels it without any doubt". This is consistent with Albretch's definition of solastalgia, because it is not only oriented towards the past but in some circumstances, it "may seek its alleviation in a future that has to be designed and created" (Albrecht, 2005, p. 45). However, this does not mean that what Méndez says does not happen. In fact, it is possible to detect some signs in this direction but they are still marginal, at least now.

The second aspect I would like to note refers precisely to the response given to the "intrusion" of the Mar Menor and the role played by emotions in the whole process. As I have already pointed out, October 12, 2019 left behind a series of negative emotions connected to impotence, anger, rage, frustration and resentment because governments not responding and lack of trust in them made the affected population think there was no future. They could not see a positive horizon. This changed dramatically when the popular initiative was set up. The figure of Teresa Vicente is controversial, and many do not recognise her merits but those who got involved in the popular initiative or in Banderas Negras know very well that she brought hope back. This was very obvious in the zines, with a clear division between those who belong to these movements (and who mention, inevitably, the popular initiative) and those who don't. In any case, what they all acknowledge is the key role of emotions (both positive and negative) in their process towards activism: "Emotion drove us to action", another zine reads. And that action as a response to the intrusion is what mobilised the "people" we mentioned earlier. Only then it is possible to understand the appearance of that foreign body that is the Mar Menor, with an activism which is born from the shared emotion, is mediated not only by memory and the memories of an idealised childhood, but also by the desire to have a peaceful old age in a territory chosen because of its environmental characteristics.

Emotion, memory and desire; these are undoubtedly three recurring elements in far-right speeches which, on the one hand, appeal to nostalgia and on the other hand, to fear, to impose a reactionary agenda based on the desire for security and permanence. But again, that road is not inevitable. From the same starting point, we have seen the creation of a citizens' movement which, veering off course, has achieved an unprecedented success in Europe – which is why the approval of Law 19/2022 recognising the legal personality of the Mar Menor is revolutionary. Beyond other considerations, this is a lesson we must not forget. Those "people", that *ecological class* (Latour & Schultz, 2022), can emerge if we can put forward imaginative responses to each of the crises thrown at us by the intrusion of Gaia. In other words, there is always room for manoeuvre.

Notes

1 This study has benefited from the support of the following institutions: Ministry of Science and Innovation, through the research project "Crisis climática, salud mental y bienestar en el Antropoceno. Una aproximación desde la ontología histórica" (PID2021-124477OA-I00 funded by MCIN/AEI/10.13039/501100011033, by "ERDF A way of making Europe").
2 It was beaten, by a few months, by Miguel Ángel Hernández's *Anoxia* (Hernández, 2023).
3 Falling salt levels in the Mar Menor in the last few years, as described in Chapter 1, has triggered, as expected, a decimation of the noble pen shell clam population in the lagoon as the parasite has managed to overcome the now weakened entrance barrier.
4 Margarita is the fake name under which I have put together individual conversations with many of the activists in this research who have chosen to be anonymous. Therefore, she is not a real person but a composite of several women who, at different times and in different circumstances, shared their experiences with me.
5 Let's not forget that an episode of severe weather conditions had taken place exactly a month earlier.
6 See Chapter 5 in this book.
7 At least until a year ago, when the white stain described by Julio Mas in Chapter 1, started to be visible from the Rambla del Albujón estuary.
8 With this term, a slight variation of Victor Turner's suggestion relative to "social dramas" as a way of accessing the ultimate fabric of the more or less unspoken social agreements underpinning a community, we aim to include the "environmental" agreements supporting that community in this fabric. We thus expand both its content and the political community around it, to include those nonhuman beings which, like the Mar Menor, claim its belonging to it at that dramatic time. For a more detailed definition of the term, please see (Mesa del Castillo & Zaragoza, 2025).
9 Despite being a course available to the general public, several associations announced it to their members. This meant that nearly 70% of the attendees were eventually connected to an association.

Reference list

Albrecht, G. (2005). «Solastalgia» A New Concept in Health and Identity. *Philosophy, Activism, Nature*, *3*, 41–55.

Albrecht, G. (2011). Chronic Environmental Change: Emerging 'Psychoterratic' Syndromes. En I. Weissbecker (Ed.), *Climate Change and Human Well-Being: Global Challenges and Opportunities* (pp. 43–56). Springer New York. https://doi.org/10.1007/978-1-4419-9742-5_3.

Albrecht, G. (2019). *Earth Emotions: New Words for a New World*. Cornell University Press.

Albrecht, G., Sartore, G.-M., Connor, L., Higginbotham, N., Freeman, S., Kelly, B., Stain, H., Tonna, A., & Pollard, G. (2007). Solastalgia: The Distress Caused by Environmental Change. *Australasian Psychiatry, 15*(1_suppl), S95–S98. https://doi.org/10.1080/10398560701701288.

Alcaraz León, M. J. (2022). On the Aesthetic Appreciation of Damaged Environments. *The Journal of Aesthetics and Art Criticism, 80*(4), 420–431. https://doi.org/10.1093/jaac/kpac037.

Belando-Torrentes, M. D., García-Muñoz, M. d. R., Ramos-Segura, A., Franco-Navarro, I. J., García-Moreno, P., & Ruiz-Fernández, J. M. (2015). *Distribución y abundancia de las praderas de macrófitos bentónicos y las poblaciones de Nacra (Pinna nobilis) en el* Mar *Menor*. https://digital.csic.es/handle/10261/320947.

Biel, J., & Brent, B. (2022). *Make a Zine!: Start Your Own Underground Publishing Revolution*. Microcosm Publishing.

Hernández, M. Á. (2023). *Anoxia*. Anagrama.

Hwong, A. R., Wang, M., Khan, H., Chagwedera, D. N., Grzenda, A., Doty, B., Benton, T., Alpert, J., Clarke, D., & Compton, W. M. (2022). Climate Change and Mental Health Research Methods, Gaps, and Priorities: A Scoping Review. *The Lancet Planetary Health, 6*(3), e281–e291. https://doi.org/10.1016/S2542-5196(22)00012-2.

Lapoujade, D. (2021). *The Lesser Existences: Étienne Souriau, an Aesthetics for the Virtual* (E. Beranek, Trad.). University of Minnesota Press.

Latour, B., & Schultz, N. (2022). *Mémo sur la nouvelle classe écologique: Comment faire émerger une classe écologique consciente et fière d'elle-même*. Les Empêcheurs de penser en rond.

Méndez, B. (2023). *Lodo*. Lengua de Trapo.

Mesa del Castillo, M., & Zaragoza, J. M. (2025). Collecting Your Own Past: Bottom-up Museums and the Moral Economy of Collections. En M. Buchczyk, M.Fonck, T. Palaić, & T. Usón (Eds.), *Unearthings. Temporalities and Ethics of Traces in Collections*, UCL Press.

Mitchell, D. (2002a). Cultural Landscapes: The Dialectical Landscape – Recent Landscape Research in Human Geography. *Progress in Human Geography, 26*(3), 381–389. https://doi.org/10.1191/0309132502ph376pr.

Mitchell, W. J. T. (2002b). *Landscape and Power* (2.a ed.). University of chicago Press.

Ortín, F. (2020, noviembre 27). *Historia olvidada de la mayor catástrofe ambiental del Mediterráneo*. Ethic. https://ethic.es/2020/11/portman-historia-olvidada-de-la-mayor-catastrofe-ambiental-del-mediterraneo/.

Rivas, P. (2023, agosto 11). *El abrazo al Mar Menor se contagia por ecosistemas de toda la península en plena sequía*. www.elsaltodiario.com, https://www.elsaltodiario.com/biodiversidad/abrazo-mar-menor-contagia-ecosistemas--peninsula-sequia.

Rohan, N. (2023). *The Power of Forest Bathing: A Comprehensive Guide to Shinrin-Yoku, the Japanese Practice of Healing and Mindfulness in Nature*. Teilingen Press.

Salazar Ortuño, E., & Vicente Giménez, T. (2022). La iniciativa legislativa popular para el reconocimiento de personalidad jurídica y derechos propios al Mar Menor y su cuenca. *Revista Catalana de Dret Ambiental, 13*(1), Article 1. https://raco.cat/index.php/rcda/article/view/404058.

Schama, S. (1996). *Landscape and Memory*. Random House.

Stengers, I. (2015). *In Catastrophic Times: Resisting the Coming Barbarism*. Open Humanities Press. https://www.openhumanitiespress.org/books/titles/in-catastrophic-times/.

Tilley, C. (1994). *A Phenomenology of Landscape: Places, Paths, and Monuments*. Berg.

Vicente Giménez, T., & Salazar Ortuño, E. (2022). Los derechos de la naturaleza y la ciudadanía: El caso del Mar Menor. *Revista Murciana de Antropología, 29*, Article 29. https://doi.org/10.6018/rmu.524761.

Wylie, J. (2009). Landscape, Absence and the Geographies of Love. *Transactions of the Institute of British Geographers, 34*(3), 275–289. https://doi.org/10.1111/j.1475-5661.2009.00351.x.

Zaragoza, J. M. (2021). Jugar con cajas de arena: La constitución de la infancia como sujeto frágil. *Contranarrativas, 2*, 118–139.

Zaragoza, J. M. (2024). Mental Health, Subjective Experiences and Environmental Change. *Medical Humanities,* 50, 417–420. https://doi.org/10.1136/medhum-2023-012879.

Index

Note: **Bold** page numbers refer to tables; *italic* page numbers refer to figures and page numbers followed by "n" denote endnotes.

For Product Safety Concerns and Information please contact our EU representative GPSR@taylorandfrancis.com Taylor & Francis Verlag GmbH, Kaufingerstraße 24, 80331 München, Germany

Batch number: 10397794

Printed by Printforce, the Netherlands